RAISING THE BAR

THE FUTURE OF FINE CHOCOLATE

Our family of Fine Chocolate

ecolechocolat.com

chocomap.com

finechocolateindustry.org

RAISING THE BAR
THE FUTURE OF FINE CHOCOLATE

BY PAM WILLIAMS & JIM EBER

Wilmor Publishing Corporation

Wilmor Publishing Corporation
100 – 2023 West 4th Avenue, Vancouver, BC

Distributed by Chelsea Green Publishing

Printed in the United States of America
by Thomson-Shore Inc.
First Edition, 2012

Library and Archives Canada Cataloguing in Publication

Williams, Pam
Raising the Bar: The Future of Fine Chocolate/
Pam Williams, Jim Eber.
Also issued in electronic format.

ISBN: 978-0-9691921-2-1 - Print
ISBN: 978-0-9691921-3-8 - eBook

1. Chocolate. 2. Chocolate processing. 3. Chocolate industry.
I. Eber, Jim, 1967- II. Title.

TP640.W54 2012   664'.5   C2012-905579-4

For future generations,
let there be the finest chocolate in the world.

CONTENTS

RAISING THE BAR
THE FUTURE OF FINE CHOCOLATE

NOTICE

PERSONS attempting to find a motive in this narrative will be prosecuted; persons attempting to find a moral in it will be banished; persons attempting to find a plot in it will be shot.

Those, of course, are not our words. They are Mark Twain's, specifically the satirical notice he placed at the start of *The Adventures of Huckleberry Finn.*

Not only are Twain's words not our own, they in no way reflect our objectives or what we want our readers to do. Twain hoped his words would get people first and foremost to enjoy the story and not search for motives, morals, and plots. We hope you enjoy our narrative, too, but we have motives, morals, and plots, for sure—specifically the preservation and propagation of fine flavor cacao and chocolate—now and for future generations.

So why use Twain's words to start the book? Because Twain's notice claiming a lack of seriousness in *The Adventures of Huckleberry Finn* has the exact opposite effect: it makes you take the book seriously.

Like Twain's notice—and like fine flavor chocolate itself—this book is a mixture of seriousness and fun. In writing this book, we strived to remember the words of Charles Schultz, creator of the beloved *Peanuts* comic strip: "All you need is love. But a little chocolate now and then doesn't hurt." We may take the future of fine chocolate very seriously in this book, but we never forget that this is a book about the simple pleasures found in chocolate in its richest, most complex forms.

That said, people looking for motives, morals, and plots behind the names found in this book can stop. No, really. Stop. Seriously.

We cast a wide net to ensure a global view, but to keep things manageable for readers, we limited ourselves to a representative sample of topics and only a few perspectives on them from each country or region. There are many more people researching and growing wonderful cacao, manufacturing brilliant chocolate, and transforming that chocolate into magnificent bonbons than we could possibly include in this book. Someone could easily interview an entirely different lot—or choose to talk to people from just a few regions or a single origin—and come up with something just as tasty. We support and hope to raise the bar for all them!

PART ONE

Seeds of Change:
Genetics and Flavor

We are drowning in information, while starving for wisdom.
The world henceforth will be run by synthesizers, people able to
put together the right information at the right time, think
critically about it, and make important choices wisely.
—E. O. Wilson, *The Unity of Knowledge*

Twenty Pounds of Cocaine

That's what the white powder on Brett Beach's desk looked like: twenty pounds of cocaine. Of course, he knew it couldn't be cocaine. Dr. Dapeng Zhang wouldn't ask him to transport twenty pounds of cocaine halfway around the world to Madagascar . . . would he? Sure, Dr. Zhang had never said *what* Brett would pack and transport the samples in, but UPS, not the FBI, had delivered the package to his San Francisco office. It sure did look like cocaine, though. Then there was the matter of the six-by-nine-inch plastic bags for holding the powder and samples. Isn't that how cocaine gets packaged? What would the customs officials in Africa think? Of course, cocaine usually travels *into* the United States . . .

"This is so bad," Brett thought. "I'm going to show up in a country that had a coup for two years carrying twenty pounds of white powder I can't completely explain." In the movie version of his life, this would be the moment things go terribly, horribly wrong.

A call to Dr. Zhang explained the sampling procedure, but Brett still had plenty of time to think about what *could* happen: The trip to Madagascar from San Francisco is one of the longest trips in the world—a flight to Washington, DC, another flight to Dakar, a connecting flight to Johannesburg, and finally, a puddle jumper to Madagascar's capital, Antananarivo. Twenty-six hours of flying over two days—which Brett likens to a bad college hangover multiplied by five but without any of the fun.

And that's just the first part of the trip. To get to the sampling area, Brett would then fly north to an island off of

Madagascar and haggle with the locals (all of them trying to charge him more because he is foreign and has a bag he'd prefer not to open) to take him back on a boat to the mainland. Back on the mainland, he'd take a taxi into Ambanja, a town not unlike the Wild West: one bank, one post office, and Western Union. From Ambanja, the end is in sight—just a long hike through the world's most densely populated jungle of endemic plants and animals to get samples of one of the most treasured and increasingly rare substances in the world.

If you think all this sounds less like a book on chocolate and more like a treatment for a movie in which an evil genius hunts the jungle for a rare ingredient to fuel his plot to bring the world to its knees, you're right. But it's not. For the record, Brett Beach is not an unsuspecting drug courier; he is a partner in Madécasse Chocolate. Dr. Zhang is not Dr. Evil; he is the lead research geneticist at the USDA's Agricultural Research Service (USDA-ARS) and its Sustainable Perennial Crops Laboratory. Neither Brett nor Dr. Zhang is chasing world domination; they're chasing flavor, specifically the DNA of cacao trees and the origin of the finest chocolate in the world—chocolate that has absolutely brought more than a few of us to our knees. And the future of that chocolate has a lot to do with the leaves Brett and a few farmers in Madagascar put in that white powder.

And no, that white powder was not cocaine; this is story about cacao not coca—think FCIA (Fine Chocolate Industry Association), not CIA. But also think CSI: that white powder in question is a drying agent—a silica gel—forty grams of which are added to a plastic bag along with a fresh leaf from a cacao tree. The silica agent removes all of the moisture from the leaf within twelve hours, thus protecting the DNA from degrading before it reaches the lab. In that hot, sticky jungle of Madagascar in early 2011, Brett Beach was pursuing genotypic identification and its connection to fine flavor cacao. He was gathering samples on behalf of the FCIA and the USDA-ARS to help map that world of cacao flavor.

Again, this is no movie, but the first part of our genetic story of the future.

Tasty Genes:
Flavor and Theobroma Cacao Germplasm

In 2008, three years ahead of schedule, the people and organizations, including Mars Inc. (which fully funded the project) and the USDA-ARS, who had successfully mapped 92 percent of the 3,500 genes in cacao germplasm, decided to release the results of the sequenced genotype (Matina 1-6) to the public and created the Cacao Genome Database (CGD). The decision by the CGD collaborators to go public generated excitement in and around the industry. The story made the national news.

In its first three years online, the CGD website (http://www.cacaogenomedb.org/) received more than 16,000 unique visitors from 131 countries. Not much for a site that blogs about Brangelina, but genetics? This was a landmark moment in the history of cacao in general and a turning point for discussions about fine flavor cacao in particular.

The CGD release was what many scientists and researchers call an essential new tool in the cacao toolbox. It unleashed a flood of new ideas and research as well as additional genotype maps from Mars, among others. Inevitably, a small amount of the work involved genetic modification and "better" cacao through chemistry. Most of the genetic research, however, centered on improved breeding practices, disease resistance, productivity, and genetic identification of the beans and flavor, and quality figured prominently in many of those conversations. For sure, genetic "fingerprinting" of trees had been available for years before the CGD, and many growers used genotyping to identify their cacao. But that genetic identification simply revealed that a bean was Amelonado Number 2—not whether it tasted good. Now people were asking at the genetic level if it had fine flavor potential.

Databases had existed for decades with information that might have been useful in connecting genotype and flavor. But Gary Guittard of the Guittard Chocolate Company, the oldest family-owned and operated chocolate company in the United States, explains that previous efforts to utilize that information and collaborate had fallen short. With twenty three-plus years of experience in genetics and flavor, he ought to know. According to Gary, "Up until a few years ago these collections were considered treasures, and they were not shared with other countries around the world. At the same time, these existing cacao collections were largely underfunded and remained incomplete or incorrect in terms of origin." Today and for the future, there is hope. Following the CGD release, Gary notes, "A new agreement has allowed these collections to be shared, and this makes them the repository for the genetic history of cultivated cocoa and others as well. That must be shared around the world now."

What's the hurry? Gary speaks for many in the industry when he says, "My fear is that down the road someone says, 'Hey, maybe some of these flavors are worth taking a look at for breeding' or whatever, and they will not be around. We need to complete a map of what these collections hold and complete the collection before many of the genetics are lost back again into the dark jungle from where they came." Hyperbole? Maybe a little, though hunting for wild cacao in Madagascar or the Upper Amazon has been likened to being in *Jurassic Park.*

Hyperbole aside, there is plenty of debate on what the future holds for fine flavor chocolate, and little to no dispute that there is incredible pressure on the world's cacao stocks. In 2012, Martin Turton of the United Kingdom's Food and Drink Federation told the BBC that by 2020 the world will need another *million tons* of cacao beans, the equivalent of another Côte d'Ivoire (the Ivory Coast, the world's largest producer of cacao) to keep pace with demand. Recent boom years in production have allowed supply to keep up, but demand shows no sign of ebbing. Most countries have witnessed increased demand for chocolate of all kinds, and every major market research firm predicts global increases in chocolate sales. The world's fastest growing market, China, is already up 40 percent since 2008 and expected to grow more than 10 percent each year for the foreseeable future. (Godiva alone plans on opening dozens of stores by 2013.) India is also coming online, not to mention huge populations in Brazil, Indonesia, and other more developed cacao-growing countries with broad and sophisticated tastes for chocolate. In the face of such worldwide demand, the pressure on all cacao stocks will only get fiercer.

Flavor trees, which are the most vulnerable to disease, environmental changes, and replacement by other crops or more productive lower-flavor hybrids, sadly account for an ever-diminishing percentage of the world's overall cacao production—

around 5 to 7 percent, depending on whom you ask (making it the bean source of about $5 to $7 billion of what Euromonitor International estimates is a $102.3 billion global chocolate industry). Something must be done to increase the production and value of flavor trees in the face of these challenges. This is not the easiest of tasks, though plenty of good ideas have been floated and even acted on to improve the overall picture: efficient and varied farm growing plans and business models; higher wages and better quality of life for farmers; better husbandry; more disease-resistant and productive hybrids; improved and innovative techniques for fermentation and drying; different manufacturing methods and possibilities; certifications; industry and consumer education on every level—the list is long, if not endless.

Acting on all of these, as we shall see, plays a part in the many possible futures for fine flavor cacao and cacao in general. But in cacao, as in life, all futures must start somewhere, and for this fine flavor story it is in understanding the flavor connection to the building blocks of those cacao beans—their cells and DNA—to better identify and differentiate them from one another. As Dr. Lyndel Meinhardt, the USDA-ARS Sustainable Perennial Crops Laboratory research leader says, "Multiple components go into the flavor of the beans, especially fine flavor beans. Look at it as a series of fourths: a fourth of your flavor could be associated with the environment, a fourth with the fermentation, a fourth with the roasting process, and a fourth with the genetics."

Simply put, flavor is one indication of a bean's genetics, and strong genetic origin has the potential to yield the best flavor. But as compatible cacao trees cross-pollinate readily, it becomes very hard to determine the genetics of a tree—or entire orchard—without DNA testing. A map that connected flavor and genetics could "speak" for that cacao and not only say what the beans are but also how they might taste, in order to help protect, sustain, and even market those fine flavor stocks for the future.

Wait: Before We Continue with Genetics, What Is the Definition of "Fine Flavor Cacao"?

In Jacobellis v. Ohio, the 1964 US Supreme Court case about defining what constitutes obscene pornographic material, Justice Potter Stewart wrote, "I shall not today attempt further to define the kinds of material I understand to be embraced within that shorthand description, and perhaps I could never succeed in intelligibly doing so. But I know it when I see it." Justice Potter's line "I know it when I see it" is often invoked when strict definitions, parameters, and subjective criteria escape us.

So let's be clear about one thing before going any further: How do we know what fine flavor cacao is? We know it when we taste it.

Truth is, there isn't a universally accepted definition, short or long, scientific or commercial, of "fine flavor cacao." We know it exists and is different from bulk or ordinary cacao. The CGD has the genome nearly sequenced, and other genetic tools and technologies are available to evaluate it. There are largely universal protocols available to appraise its flavor. The fine flavor industry works closely with scientists, breeders, and farmers to grow it. Many groups and projects have studied the factors from environment to fermentation to shipping to processing that can affect its flavor as it turns into chocolate. Artisanal manufacturers and chocolatiers tout all these characteristics, as well as origin, terroir, and various certifications, when marketing their chocolate, to confirm its quality and justify the higher prices.

And still, really, there is nothing definitive for all of us to hang our palates on.

Not that everyone thinks that is necessary right now. Joe Whinney, a former cacao bean buyer and founder of Theo Chocolate, is a big supporter and marketer of the genetic work in cacao but says bluntly, "I don't find a lot of broad relevancy to the concept of fine flavor. I understand it. But I don't think the consumer is thinking about that. They think about origin. They think about percentage a little bit more than they used to. But ultimately they are still seeing chocolate as this sweet treat that if it's dark it might be a little bit better and better for me and what kind of nuts does it have in it . . .?"

Okay, but that doesn't mean it hasn't been in the industry's interest to create some definition and make it relevant for all the industry stakeholders. "After all, if we do not talk about it here, in the industry, no one will," says Ed Seguine, who spent twenty-five years at Guittard before taking his current job as chocolate research fellow at Mars Chocolate North America, a major funder of worldwide scientific research on chocolate. And some very smart people have tried to define fine flavor—some at great expense. For example, Ed was one of a group of authors on the Project to Determine the Physical, Chemical and Organoleptic Parameters to Differentiate Between Fine and Bulk Cocoa (2006), a $1.67 million, five-year study sponsored by the International Cocoa Organization (ICCO) and prepared by the Instituto Nacional Autónomo de Investigaciones Agropecuarias (INIAP) in Quito, Ecuador.

The summary of the project proposal, which evaluated beans from Papua New Guinea, Trinidad and Tobago, Venezuela, and Ecuador, is worth quoting at length as it explains myriad reasons that such a definition was and still is so important:

> The definition of fine or flavour [*sic*] cocoa remains controversial as there is no single universally-accepted criterion that could be adopted as a basis for determining whether or not cocoa of a given origin is to be classified as fine or flavour cocoa. . . . The main objective of this project was to develop the capacity of all involved in the cocoa sector to adequately differentiate between fine and bulk cocoa, thus improving the marketing position of fine or flavour cocoa. The specific objectives of the project were to establish physical, chemical and organoleptic parameters enabling the evaluation of cocoa quality in relation to genotype and environment, and to disseminate selected parameters, methodologies, standards and instruments to be used in the evaluation of cocoa quality.

Unfortunately, the project didn't get very far. For all its detailed work on fermentation and drying trials, chemical assessments, cocoa liquor preparation and chemical analysis at different roasting temperatures, organoleptic assessment of sensory characteristics by a trained panel, and DNA profiling, the project, in its own words, produced mixed results. It produced plenty of evidence "to reinforce the scientific basis that sustained the concept that fine and bulk cocoas had specific attributes influencing the way these products were used." It also showed how genetic variability produces sensory inputs (flavors) like fruity, nutty, caramel, and acidity. It found that "the theobromine/caffeine ratio proved to have consistently good discriminating power to segregate fine or flavour from bulk cocoa," which companies like Luker Cacao in Colombia have integrated into their breeding programs.

And one of the ICCO report's observations proved particularly prescient: It not only realized the possibilities of international collaboration, but also undercut a potential barrier to cooperation—there would be little competitive disadvantage. In fact, the report stated in bold, "[I]n general fine cocoas from different origins have distinct flavour profiles, thus eliminating market competition among them because they each occupy distinct niches in the market. This situation provides a strong incentive for future collaborative research among fine cocoa producing countries to advance the development of the fine cocoa market."

Two years later, the Cacao Genome Database was released and a collaborative spirit thrived, inspiring outstanding ongoing genetic work with a focus on flavor in cacao-growing countries. Dr. Darin Sukha, a research fellow at the highly regarded Cocoa Research Unit at the University of the West Indies in Trinidad and Tobago and one of the authors of the ICCO report, is more than thrilled with the progress made on genetics and flavor for farmers, breeding programs, and the industry as a whole since the report was issued. "In the past, each chocolate company had their own protocol and vocabulary for talking about flavor. It was like two people sharing a common culture but speaking a different language. They knew what they were talking about but they couldn't really communicate, change results, or have any kind of harmonization of opinions. Now, we have marker-assisted selection with the release of the cocoa genomes. We have some degree of standardization in vocabulary and in protocols that for the first time allows broader collaboration."

This is not to say that everyone is throwing open the metaphorical doors and working together. Some origin countries, if not the growers themselves, can be very controlling of their plant materials, and the manufacturers are protective of the relationships that they have made with the growers. The Dominican Republic has literally closed its borders to protect against the spread of disease. Governments in Brazil, Peru, Colombia, and Venezuela, as well as anyone exploring the Upper Amazon basin, want to control their findings as much as they can until they can capitalize on them, which is understandable. Knowing you have great or even unique plant materials and wanting to benefit from breeding them before anyone else makes good economic sense in a market driven by marketing as much as flavor—even if the cacao flavor profiles are different from others on the market.

Yet the collaborative spirit lives. The World Bank Development Marketplace is funding a new project led by scientists at the University of British Columbia titled Identification and Promotion of Ancient Cacao Diversity through Modern Genomics Methods to Benefit Small-Scale Farmers. The ICCO is supporting a new project that will use the advances in genotyping flavor to try to find ways to identify and map quality attributes from flavor at origin to improve the capacity of countries to market their cocoa.

This same collaborative spirit also inspired the FCIA and USDA-ARS to partner in creating the Heirloom Cacao Preservation Initiative—the first-ever genotype map with a focus on flavor cacao trees. This also takes us back to where we started our story: the sampling of the cacao trees in the jungles of Madagascar. A look at how this initiative formed and proposes to move forward in the future serves as an example of how genetics can be put to use by the entire industry to identify and promote the flavor *and* genetics of cacao and use that information to aid their beans' preservation, propagation, and marketing position.

Look at the Map:
The Heirloom Cacao Preservation Initiative

Officially launched at the Fine Chocolate Industry Association conference in January 2012, the Heirloom Cacao Preservation Initiative (http://finechocolateindustry.org/hcp) is designed to connect growers, traders, chocolate makers, manufacturers, and consumers to a very specific set of goals: know where the world's finest flavor beans are; tie their flavor to their genetics; and use that information to improve and help ensure fine flavor cacao quality and diversity, and preserve, protect, and propagate fine flavor beans for future generations.

The object of the Heirloom Cacao Preservation Initiative (HCP) is to identify and classify heirloom flavor, not to replace or denigrate bulk or ordinary cacao—no genetic work on flavor could or should. Genotyping flavor won't say beans like the Forastero types that make up bulk cacao are bad and therefore all Forastero is bad. In fact, the HCP will likely show there are some very nicely flavored Forasteros. And no matter how successful any initiative—genetic or otherwise—is, fine flavor beans can't replace those used to make a Tollhouse morsel or a Hershey bar on Halloween.

Cacao will never return to the days more than a century ago when the majority of the world's beans were what we today consider to be fine flavor. The candy market is too big and reliant on lower-cost beans; costs will always be too high to make fine flavor cacao anything more than the "high end" of the market. Instead, the HCP seeks to maintain the diversity that is getting lost so that there *is* a high end in the future—so that when someone wants to spend money on the deep pleasures of a remarkable *palet d'or* or bar of fine flavor chocolate, it still *exists*. Life without that bar or bonbon that makes you scream, "Oh my God!" may seem unimaginable but it is possible, as growers continue to replace fine flavor cacao trees (and indeed entire farms and plantations) with high-yield, disease-resistant, less flavorful cacao hybrids and clones, cattle, or other crops like bananas, pineapples, oil palms, corn, and soy.

The idea for the HCP emerged in the summer of 2010 when representatives from the FCIA met with Dr. Lyndel Meinhardt and Dr. Dapeng Zhang of the USDA-ARS to talk about fermentation, DNA testing, and cacao growing. The discussion evolved into a broader discussion of flavor and DNA, and Lyndel thought the USDA-ARS lab could help the FCIA identify the cacao using the 5,300 samples in the worldwide database as baseline principal coordinates. (The USDA has been instrumental in working with several international collections and updating their accuracy.) Then, unlike so many big ideas, something actually happened. In October 2010, the FCIA brought its committee of artisan chocolate makers in on the idea and support was strong. When asked where the committee wanted to get started and learn more, Madagascar was the top choice. No surprise. Madagascar is the fourth largest island in the world, but the cacao trees are all within a twenty-five mile radius in the northwest—a small area that accounts for 1 percent of the world's cacao production and its outsize reputation in the world of fine flavor chocolate. That's how Brett Beach wound up standing in Ambanja, Madagascar, in early 2011 bearing a suitcase full of white silicone powder and plastic bags, explaining to the town's farmers how to collect leaf samples for genetic testing.

While the FCIA and USDA-ARS prepared to reveal Brett's findings at the FCIA conference in July 2011, they simultaneously began to develop the wording for the mission of the HCP, striving to make it a model for and complement to other preservation efforts on a collective and collaborative scale. In December 2011, the FCIA established a specific cooperative agreement with the USDA-ARS to develop their ideas further and announced the partnership to members at the FCIA conference in January 2012 along with an appeal for funding. And the members responded: Twenty-five large and small chocolate companies and industry stakeholders stepped up as Founding Circle members in the first half of 2012, providing the funding to allow the HCP to move forward that July. (The list of HCP Founding Circle members can be found in the back of this book.)

Now, the real work begins. The HCP seeks to engage as large a partner network as possible to identify at-risk heirloom cacao populations. Reflecting the cooperative spirit that led to the mapping of the cacao genome years before and the idea for the HCP, the FCIA calls all potential HCP partners "collaborators," be they cacao growers and processors or those utilizing cacao, such as traders, chocolate manufacturers, or artisan chocolate makers—and will reach out to them worldwide on an ongoing basis to encourage their participation in the HCP program. All of these collaborators may submit their beans for evaluation or simply inform the HCP of at-risk heirloom cacao populations in need of evaluation.

But while any beans, like those nicely flavored Forasteros, can be submitted to the HCP and evaluated for their flavor, not every bean can or will be deemed an heirloom—regardless of any pedigreed origin or even if existing genetic work on that bean says it could be fine flavor. The HCP has a strict "Standard of Identification" based on criteria and guidelines developed by industry experts to evaluate flavor and validate that a bean is indeed an heirloom. So, what does that mean? It means flavor comes first in this genetic endeavor. This "flavor first" motto is exactly why the FCIA chose the word "heirloom" and its basic Webster's definition—"a cultivar of a vegetable or fruit that is open-pollinated and is not grown widely for commercial purposes [and] often exhibits a distinctive characteristic such as superior flavor or unusual coloration"—to frame its initiative.

"First of all, it has got to taste good," says Dan Pearson, chief executive officer of Marañón Chocolate and an FCIA board member who helped develop the HCP. "If it doesn't taste good, we are not going to proceed with the genetics. Genetics alone say nothing about flavor. That's about classification and that's the second step. We need to start with flavor, not whether a bean is pure such and such or ancient so and so."

But can taste be evaluated objectively like genetics? No. And the HCP does not say otherwise. The process starts objectively: After bean samples are submitted, the HCP will anonymously process the beans in volunteer professional labs into "liquor" (what the industry calls unsweetened chocolate) and semi-sweet chocolate. The USDA-ARS will then use gas chromatography to develop and record science-based flavor profiles for those beans. But the next and decisive step is purely subjective and tasty: the submission of those chocolate samples to the HCP Tasting Panel for flavor analysis. This international panel is currently made up of seven experts with fifteen to twenty-nine years' experience in chocolate and who have all served as professional evaluators of cacao bean flavor in the past: Chloé Doutre-Roussel (Chloé Chocolat, France); Gary Guittard (Guittard Chocolate, US); Ray Major (The Hershey Company, US); Jorge Redmond (Chocolates El Rey, Venezuela); Ed Seguine (Mars Inc., US); Dr. Darin Sukha (Cocoa Research Unit, The University of the West Indies, Trinidad & Tobago); and Franz Ziegler (Ziegler Consulting Switzerland).

If the HCP Tasting Panel unanimously scores the flavor of a bean sufficiently high enough (say, ninety to one hundred if the panel uses a scale similar to one currently used in wine), only then will the bean be certified an "heirloom," genetically mapped, and preserved in the HCP Database of Heirloom Cacao. That database will be maintained with the USDA-ARS and will connect those genetics to the beans' flavor profiles. If all goes as expected, the HCP will then map all these heirloom cacao genotypes and flavors right down to their GPS location.

Think of the Heirloom Cacao Preservation Initiative and other projects just getting started as genetic stakes in the ground for flavor. In the short-term, the HCP will apply new genetic standards for flavor identification to help protect and propagate heirloom beans, mapping their flavors down to their GPS locations. In the long-term, chocolate makers, chocolate manufacturers, and chocolatiers will be able to indicate the HCP certification on their labels, supporting growers of fine flavor beans and alerting buyers to the presence of independently verified great flavor. Ed Seguine calls all this work the beginning of a "paradigm shift for the future ways we look at the flavor" so we don't "lose it like tomatoes and strawberries did." Prior to the HCP, Ed had put his company's money where his mouth is to genotype relic cacao in Trinidad and Tobago for the World Bank Development Marketplace. One project, called Modern Genomics Methods Benefiting Small Farmers' Value Chain in Trinidad and Tobago, certainly is a mouthful but at least it had taste thanks to Ed: Initially planned as a collection of cacao for genetic typing, Ed expanded the funding to include flavor evaluations on the samples. He used Mars to process the majority of the cacao samples into chocolate so their larger value could start to be understood by the farmers.

Ed calls this "education from an engaged standpoint, where everybody gets involved on this journey of discovery." And of course, that is the key to all these efforts and the most imperative step in preserving diversity: education on identification and issues around at-risk heirloom cacao, not only for the fine chocolate industry but also for consumers and media. We must explain how this type of knowledge and data will add value and make a real difference in the future. Because that's the end game: explaining and reinforcing the importance of genetics and genotyping to flavor in order to preserve, protect, propagate, and identify more flavor beans in the future.

It should work—at least on paper: The projects touch on all the buzzwords most people in the fine chocolate industry and their customers care about: biodiversity, sustainability, and endangered species, to name a few. But no one is working on better tasting paper. If we could taste chocolate on paper, explaining what this all means for the future would be a little easier. That said, if you haven't already, we recommend unwrapping your favorite fine flavor bar or opening a box of bonbons to make this next part a lot tastier and to remind you of what is at stake.

Trying to Explain What All This Means

Albert Einstein said, "The difference between what the most and the least learned people know is inexpressibly trivial in relation to that which is unknown." Genetics and genotyping flavor in cacao falls between the known and unknown for people who live for and love fine chocolate. Even to the most educated and informed people in the business, genotyping flavor can be as messy and difficult to grasp as a chocolate bar on a hot summer day. It's not so much scientific gobbledygook as an abstract thing to give a damn about. The FCIA professionals who have experience up and down the chocolate food chain, and who offer deep insights into every social and production issue in the industry, are still just as likely as the rest of us to call chocolate yummy. Many of them just wish that modern manufacturing focused on fine flavor as it did years ago. That's the reason Steve De Vries of De Vries Chocolate proudly says his company is "one hundred years behind the times."

The industry is one hundred years behind the times in another way too. Cacao is an ancient tree but for all the amazing genetic, scientific, technological, and other advances, chocolate production in places like Madagascar hasn't changed much since the French started harvesting it in 1903. So while millions and millions of us speak the "language" of chocolate and 6.5 million farmers—many for generations—make their living growing cacao, the genetics of cacao is like a modern dialect few can speak yet. It is undoubtedly an essential part of the future, but understanding and processing the torrent of information and the possibilities it offers is overwhelming and, honestly, never going to make you drool with anticipation the way unwrapping a bar or opening a box of bonbons will.

Simply put, genotyping and understanding the genetics of fine flavor cacao will lead us to anticipate and better appreciate that fine flavor. But explaining genotyping and the genetics of fine flavor cacao—making them interesting or at least top of mind so that people who have a stake in the future of fine flavor cacao buy into their importance for their beans, businesses, and customers in the future—is a process. And if you think explaining this process is challenging here, imagine doing it in Madagascar. Not easy, even if like Brett Beach you speak Malagasy, thanks to six years in the Peace Corps and international development projects. He had to explain not just the sampling, but why genotyping flavor matters for the future, and how it is going to bring value back to the farmers.

"The farmers got it as much as I could convey it," Brett says. "This is a challenging and abstract subject to distill, let alone translate to farmers, especially if you don't have much of a clue what genetics is in English." The same could be said about the people on the other side of the world who gathered at the FCIA conference in July 2011 to hear the results of the testing on Brett's samples; like the farmers in Madagascar, they got it as much as "it" could be conveyed.

The FCIA chose the right person to convey the information: Dan Pearson, and not just because of his work at Marañón Chocolate or that he helped found the HCP Initiative.

Dan has the kind of deep-yet-cheerful voice that makes you lean forward and want to buy into whatever he is saying. Dan also understood on another level the meaning of the results he was about to deliver: He had already been a newsworthy beneficiary of genetic identification.

In 2009, the USDA-ARS tested a sample of the pure white beans that Dan and his partner, Brian Horsley, had stumbled upon on the Fortunato Farm in the Marañón Canyon of Peru. Dr. Lyndel Meinhardt compared them to the existing genetic database and confirmed that the beans were Pure Nacional cacao—a variety thought to have disappeared in 1919. In January 2011, a *New York Times* food writer tasted Marañón's chocolate and started writing a great review when she became as interested in his discovery as the chocolate. The result was the first national newspaper story to link flavor and genetics for consumers. Lyndel was quoted in the article explaining how Marañón's white beans were genetic mutations that happen when trees are left undisturbed for hundreds of years. As a result, they have "fewer bitter anthocyanins, producing a more mellow-tasting, less acidic chocolate." Anthocyanins? This was not "Science Times." This was in the *food section*.

Thus Dan already knew what it felt like to stand between Einstein's knowns and unknowns when it comes to flavor and the genetics of cacao. He sums it up perfectly when recalling the conversation he had with Brian immediately after learning the USDA results.

"My God, do you know what this means?" he said elatedly.

"Do *you*?" Brian replied.

"No, I don't!"

And that's exactly the déjà vu feeling Dan got two years later as he revealed the initial results of Brett's Madagascar samples at that FCIA meeting in July 2011. Standing before a slide with the eighteen samples plotted as red triangle coordinates on a graph with the four established cluster names for classifying cacao germplasm—Upper Amazon Forastero, Trinitario, Ancient Criollo, and Amelonado—in their own quadrants, Dan's voice suited the dramatic results: Two samples were pure and rare Ancient Criollo—the first the USDA-ARS had put in the database; six of Brett's other samples were genetically pure Amelonado, another very rare species that is disappearing; and eight of the remaining ten samples were flavor beans in the Trinitario cluster.

These results would be extraordinary under any circumstance but especially for the initial efforts of the FCIA/USDA-ARS partnership. There was a perceptible energy as people realized eight of the eighteen samples were extremely rare and alive—essential in cacao because to guarantee the propagation of a tree's DNA, you cannot grow trees from seed; you must clone or graft from the tree itself. Cacao flowers are promiscuous and will mate with any other cacao pollen that comes along; their "seeds" can easily not have the same genetics as the tree.

"The scientists," Dan said, "were extremely excited and your farmers will be, too. We didn't mean to surprise you, but we did."

And then . . .

At moments like this in the movies, the audience stands up and applauds—that's how remarkable these results were in the world of fine chocolate. But in the real world, like that night, profound moments like these often are punctuated by silence as people process the information—just like the guest of honor at a surprise party. Only the FCIA audience wasn't sure what this party was all about. Surprise! They went from processing the facts to arriving at the same "I don't know what it means" that Dan did when he first heard the news about his Pure Nacional. Any oohing and ahhing was reserved for the chocolate tasting that followed the presentation.

As Dan said later about the reaction, "Quite honestly I didn't give the silence that much thought. Their reaction was probably the same as mine when I first found out, which was, 'Wow, this is significant, but I don't know how significant.'"

Many people in the FCIA audience that night and across the fine flavor industry in the months that followed were also probably wondering the opposite: "If the industry has gotten this far without a strict definition of fine flavor, why do we need to genotype it? Who would want this information?" But those are the wrong questions to ask. The right ones are: "Why wouldn't everyone want it? Why wouldn't farmers who put hard labor into their cacao trees want to know more precisely what they have? Why wouldn't that information be of interest to their customers and anyone else? People are willing to pay more for food with deep connections to origin. So why not offer even more information by identifying the beans and classifying them in terms of flavor to add value to the farmers' product?"

The importance of this is not lost on the researchers, either. "Creating an awareness of the value of this material, we automatically add value to it," says Dr. Darin Sukha of the Cocoa Research Unit at the University of the West Indies in Trinidad and Tobago. "Look at Chuao. Who cares about Chuao outside of the broad context of cocoa? Is it perceived? Is it real? A fair amount of both. You have a lot of hype created by marketing but now you balance it with information that shows it can deliver on a genetic [promise]. Having a protocol for assessing flavor has been the first step in now critically examining cacao flavor the same way that you would do for wine and understanding what the components are that affect flavor."

Dr. Lyndel Meinhardt of the USDA-ARS seconds that: "The genetic genotyping of the trees raises a whole other level of understanding for the chocolate makers. And as consumers get to know fine flavor chocolate, they are starting to move more toward it. That has raised the level of the discussion to 'Am I really getting what I think I'm getting? Is there something even better?' as growers and manufacturers try to differentiate their product from everyone else's."

Yes, even in the face of genotyping, notes Jeffrey Stern, a consultant and chocolatier at Stern Chocolates in Ecuador, some growers and manufacturers will still call their beans an Arriba variety, even though that is not a germplasm cluster and there is debate if there is such a thing as Arriba flavor. But at least with a genetic map of flavor, manufacturers and consumers will have new information to ask pointed questions and force more clarity in the world of fine chocolate. They can ask, Jeff says, about something labeled Arriba for "traceability of both the bean variety and geographical origin for the beans used in the chocolate" and have some clue about what that means.

In an age when consumers care so much about the origin of their food and what their food is made of, genetic identification can go beyond marketing and ensure that a cacao is what a manufacturer expects, and that what's inside the package is what the manufacturer says. Imagine a world where listing germplasm clusters on fine chocolate packaging becomes as prevalent as country of origin and other identifiers from percentage to single origin to fair trade and organic certifications—and correlates to what's inside. To many consumers, percentage remains an indication of quality, not simply cocoa mass, and with no regulations or standards to tell them otherwise, even genetic identification won't reveal anything more about that percentage. For example, something could still be labeled a "65% Madagascar Criollo" bar with only 1 percent of that bar being from that origin and cluster. Now imagine a world in which consumers buy a dark chocolate bar from Madagascar that instead of just saying "80% cacao" says "30% Criollo/50% Trinitario" from the origin—and understand what it means, the same way most drinkers of wine understand the taste of a cabernet/merlot blend from Napa.

Joe Whinney of Theo Chocolate would welcome this as part of an industry-wide push for more transparency. One of the first manufacturers to blog about cacao genetics, Theo has used genetic fingerprinting for years to guarantee ingredients through their supply chain from harvest to delivery. This transparency is a huge part of Theo's value proposition, and the genetic fine flavor component could be an important addition. Dan Pearson agrees and hopes in the next five years that genetic identification can do for honesty in the packaging of fine chocolate what Felchlin, Valrhona, and other big fine chocolate makers did years ago with ingredients and origin. When those companies started saying what was really in their chocolate and revealed the source of their beans, they forced the hand of mass-market chocolate makers who were selling cheap quality for high prices. (This story is nicely captured in Mort Rosenblum's *Chocolate: A Bittersweet Saga of Dark and Light*.)

"There has been plenty of 'fine chocolate' that has worked in the marketplace. But it is fine chocolate because the consumer was told it was; the emperor has no clothes," says Gary Guittard. Better not to ask any questions, because if you don't ask questions you won't get any answers. Genetics allows informed customers and consumers to ask those questions: "What are the beans in that expensive chocolate bar? What is it in that sixty-five percent Madagascar bar? Hmmmm . . . this is good but is it really made from the wild Bolivian cacao? Are those beans really from Chuao? How much of it is from origin?" These are the questions about the packaging of fine chocolate that could be asked and understood in the future once genotyping flavor becomes a wider part of the cacao conversation, resulting in a new understanding, if not definition, of fine flavor chocolate.

Of course, genotyping flavor beans only guarantees what you *could* have in terms of flavor in that 65 percent, not how it will turn out given that so many other variables from environment to terroir affect flavor even before the fruit is harvested. Genotyping and percentages alone can't tell you if that Ancient Criollo has been handled well. Pitfalls are numerous after harvest in fermentation and drying—and still you are a long way from making chocolate. As Joe Whinney says and many manufacturers echo: "You know there is a straight line between a merlot grape and how it might taste when interpreted by a winemaker. Cacao doesn't have that straight line. The opportunities to impact the flavor of chocolate by the manufacturer are so much greater. My interpretation of that bean might be wildly different from how another manufacturer might interpret it. And as a consumer, I wonder if this genetic information is useful or more confusing."

Anyone who has read this far or simply follows a thread on industry sites and social media feeds from Ecole Chocolat, C-spot, or Chocolate Life knows the genetic information needle tips toward confusing right now—fascinating, but confusing. But in the end, you can't explain—let alone preserve, protect, or propagate—flavor without knowing and seeing that flavor at its most basic level. Creating genotypic maps of flavor based on the germplasm clusters would be as basic as you get.

Interestingly, consumers and even many of us in the industry can learn much from the farmers in our identification efforts. Thinking back to the FCIA's announcement of the results of the Madagascar samples, Brett Beach disagreed with only one part of Dan Pearson's assessment: his farmers would not be too surprised at the results. They knew what most of the trees were and got both the Ancient Criollo right. Despite being from an undeveloped town in a relatively poor country, Brett says, "They knew as much as any foreign technician that works with cocoa. They knew based on the thickness and color of the leaves whether it was Criollo, Trinitario, or Forastero, and how to mark them for future reference. They just pulled the leaves off and put them in the bag. And when they marked the trees? They 'tattooed' them by tracing around the bark, cutting it off, and carving in roman numerals into the trunk."

In fact, the hard part of explaining genotyping was never teaching people the established cluster names for classifying cacao germplasm: Trinitario, Forastero, Criollo, and Amelonado. Those four clusters have been in place for more than eighty years and are commonly known as types of beans by professional and consumer "chocolate nerds," as Rick Mast of Mast Brothers Chocolate calls us. That's also how the farmers who led Brett Beach through the jungles of Madagascar knew what they were looking for.

Given how much cacao is still waiting to be found and rediscovered and mapped with the thousands of beans we do know, we'll likely need these farmers' help tattooing even more trees in the future. We'll just have to teach them some names for the new clusters. That's right—for everyone just getting their heads around Trinitario, Forastero, Criollo, and Amelonado, get ready for some new, head-exploding information in the future: a revised set of ten-plus clusters, and potentially more as the research continues, currently sampled cacaos are classified, and new cacaos are discovered. For such a relatively simply genome, chocolate is an incredibly complex food. Chocolate alone—before adding any sugar or inclusions (such as nuts)—possesses hundreds of flavor compounds (three times that of red wine). It is impractical to use only four clusters to classify the flavor possibilities. And while undoing those eighty years of accepted cluster classifications of cacao germplasm won't be easy, the ten-plus will be a whole lot more representative of the genetic diversity of cacao in the long term.

Reclassified Information

The ten new proposed clusters were first identified and named in a 2008 study led by Juan Motamayor of USDA-ARS and titled Geographic and Genetic Population Differentiation of the Amazonian Chocolate Tree. Funded by the USDA and Mars Inc. and authored by representatives from those organizations,

INIAP (Ecuador), CIRAD (Centre de Coopération Internationale en Recherche Agronomique pour le Développement, France), and CEPLAC/SUPOR (Executiva do Plano da Lavoura Cacaueira, Brazil), the report focused on improving cacao by better understanding the origin, classification, and population of the *Theobroma cacao L.* germplasm. The study's particular focus was the labeling errors that existed in the databases—errors that hindered breeding as well as diversity and differentiation analyses.

It is not too hard to follow the report's topline descriptions of the research and methods, the accompanying graphics illustrating the results, and overall conclusions. But reading the entire report is not recommended for those who don't find pleasure (let alone flavor) in Bayesian statistical analyses and devouring scientific sentences from the seemingly simple ("with the pattern of differentiation of the populations studied supporting the palaeoarches hypothesis of species diversification") to the magnificently complex ("The overall Fst value [after 1000 bootstraps over the retained loci] was 0.46 [99% Confidence Interval: 0.44–0.49]"). So let us sum it up for you: Things are a mess in cacao genetic differentiation—lots of sampling, lots of errors, little reliable data for interpretation and use, and for a genetically varied species like cacao, not enough diversity in the current cluster classifications to understand and manage that diversity. So they decided to change those classifications.

Now, much as judges do not take overturning precedent lightly, the authors of this project were thorough in their deliberations. They worked together to reevaluate all the morphological data from the International Cacao Germplasm

Database. They genotyped 1,241 existing samples in the database, and after discarding 406 that were mislabeled or highly homozygous (which basically means genetically identical, not uncommon with natural hybrids like cacao), they analyzed the 735 remaining samples. What they found was that the two traditional main genetic groups, Criollo and Forastero, and the remaining three groups, Trinitario (a Criollo/Forastero hybrid), Nacional, and Amelonado—were not cutting it for classifying cacao's diversity. They instead proposed a "new classification of cacao germplasm into 10 major clusters, or groups: Marañón, Curaray, Criollo, Iquitos, Nanay, Contamana, Amelonado, Purús, Nacional, and Guiana. This new classification reflects more accurately the genetic diversity now available for breeders, rather than the traditional classification as Criollo, Forastero or Trinitario."

The authors encouraged germplasm curators and geneticists to use their new classification to "conserve, manage and exploit the cacao genetic resources" in breeding and beyond. They then published the study with open access, placing all the information in the public domain where anyone can use it. The Heirloom Cacao Preservation Initiative, for one, will genotype flavor using these clusters and has the flexibility to adapt as more are proposed or names are changed. And they likely will. After all, not everyone is rushing to accept the new classification.

For example, no one seems upset about the loss of the Forastero cluster, which was the most undifferentiated, but some people in Trinidad and Tobago are very passionate about the loss of the Trinitario cluster even if they understand the scientific reasoning. Speaking for several local stakeholders, Dr. Darin Sukha acknowledges that the exclusion of Trinitario from the

proposed clusters was not deliberate in any way. Trinitario is a
cultivar and the Motamayor study evaluated wild types of cacao
and "original" germplasm. (Trinitario, Nacional, and Amelonado
are all traditional cultivars, but only Trinitario is a hybrid and
not wild.) But, Darin explains, "There are two sides to this
question: The academic side that really does not focus on
Trinitario as a wild grouping in this ten-plus cluster development,
and the other side that must consider Trinitario in the context of
marketing fine or flavor origins and Trinidad and Tobago as the
'birthplace' of this hybrid that is highly regarded in the fine or
flavor cocoa market."

Indeed, Trinitarios, which make up about 10 percent of the
beans in the four-cluster classification, have been grown on
Trinidad and Tobago for 200 years and those beans, along with
their clones (like Ocumare from Venezuela), are highly regarded
for their flavor (although the high-yielding Trinitario clone CCN-
51 is not, and serves as a kind of bugaboo when it comes to
discussions of hybridization and flavor). Trinitario types are a
point of national pride for Darin and his colleagues: "We as a
country should make sure that this information is not lost, but
rather seek to use it to market our cocoa. The use of the name
Trinitario in marketing of fine flavor cocoa is very important. We
should all get emotional and passionate and make sure that
Trinitario as a marketed fine or flavor origin or type is not lost,
especially for Trinidad and Tobago. Niche marketing, branding,
and geographic indicators are tools that we should exploit to
improve the marketing competitiveness of fine or flavor origins."

National pride and marketing, naming confusion, people's
natural stubborn resistance to change—all this and more means
the discussion is far from over. These clusters are a ways away

from being fully defined and gaining acceptance in the industry, let alone by consumers. After all, even experts in the field, like Dr. Lyndel Meinhardt, are just starting to appreciate the genetic diversity of this relatively simple genome: "The genetic diversity surprises me. I just don't know that we have identified all the populations of cacao that are out there and whether we have looked at all the best and finest flavored chocolates that are out there. There's still a lot of wild cacao in the Amazon River basin; we have just discovered a few new populations there that no work has been done on yet. As old as cacao production is, we have such a limited grasp of what we are working with. Cacao is so confused with the naming of it. It is so confused with how it's grown and spread around the world. It is poorly tracked from its origin. It's amazing to me that as old as cacao farming is, we know so little."

Given this, it is perfectly logical to assume another group of academics and scientists may try to rewrite the clusters again in the future as the debate goes on and more cacaos are discovered and evaluated. As a result, the new clusters are even further away from capturing the public imagination—let alone generating national news stories the way the mapping of the genome and the discovery of Dan Pearson's rare cacao bean did. "Right now," Dan says, "all this is incredibly interesting from a scientific point of view, but show me someone who actually gives a damn. Whether there are three categories or ten categories is immaterial right now. You can't chew on scientific accuracy, so it doesn't make any difference to the person eating the chocolate, even if it makes a lot of difference to the scientists."

"At least not yet," Dan adds. "If someone can take all the scientific gaga and translate it so people can learn to do more of what we did in twenty-one months, and can coax the flavor out of a white bean that takes two to three days to ferment and a dark bean that takes five to six days, we've got something. Until now, we've just been paying all these people to write these remarkable reports that nobody reads. But there's a lot of good stuff in there. Once someone takes one of these genetic varieties and learns how to bring out its flavor profile—and has a big enough marketing budget to tell the world about it? Then it becomes even more significant. Once someone with the genius and budget discovers a new variety with a new taste? Oh, you bet it is significant."

Joe Whinney agrees. "The reason we have an interest in genetics is that we like to know a little bit about what kind of flavors to expect and what kind of nutritional value they might have. I think that the industry would serve consumers better if the industry understood and respected the genetic uniqueness of cocoa more. When we just try to narrow down the three varieties that is so misleading. There are more than 14,000 known varieties of cocoa beans. Without set standards the consumers can rely on or that the industry can help manage, it can be misleading and ultimately it is not a service to the consumers or the market. I think more manufacturers might even use it to extract some value out of regions that previously wouldn't be considered that valuable. We could take some beans from Côte d'Ivoire and make some delicious chocolate with it. One country doesn't produce one million tons of crap every year. The more transparent we can be about the breadth of this flavor, the more we're actually telling the whole story, and people will have to tell the whole story behind their products right down to the genetics. Those are the stories that enrich the consumer experience." One needs only to walk into any of Patrick Roger's stores in Paris to confirm the breadth of this flavor story: dozens of bars stacked neatly on the wall identified only by their country of origin, including Côte d'Ivoire.

That's the real value genotyping flavor has the potential to add, when it comes to flavor—it tells a deeper story about the bean that adds value to the end product. It allows the industry to drill down from the clusters and get more specific—specificity that will eventually be in everybody's interest. The task may be daunting but filled with promise. If mapping flavor simply helps market fine chocolate and justify the higher prices fine chocolate commands *and* thus makes customers more willing to pay a premium for it, then more people will grow it in the future. For example, genotyping efforts linked with flavor like the FCIA's Heirloom Cacao Preservation Initiative could help farmers in Ecuador, who currently receive the same price from commodity brokers whether they deliver fine flavor or bulk beans, to differentiate those beans and find a direct buyer.

"With help from initiatives like that of the HCP, government funding, and assistance from well-respected industry veterans and farmers like Samuel Von Rutte, small farmers in Ecuador can learn to better understand and improve the flavor profile of their beans," says Jeffrey Stern of Stern Chocolates in Ecuador. "Cooperatives and small farmers can then learn to link directly to foreign buyers who are willing to pay substantial premiums for increasingly scarce fine flavor beans. Growers can also benefit from the HCP and other third-party verifications that attest to the quality of the beans after fermentation and drying and upon shipping. This process can help establish trust between direct buyers and sellers and remove the need for intermediaries. The premiums delivered to small farmers and cooperatives will then go directly to them, undiluted by intermediaries that were unwilling to pay a premium in the first place."

In Peru, Dan Pearson is already banking on the significance of genotypic identification from the propagation side. From his one Pure Nacional tree, Dan took fifty starts and by 2011 had 2,500 pure trees growing—and could have 250,000 if that's what Marañón chooses to do. Recognizing that they could not compete with Côte d'Ivoire and needed to grow more premium flavor varieties, many of the growers in Hawaii have done genetic testing since 2009 to help plant better material, grow the industry, and market it as an origin of value. In Java, Frederick Schilling, a partner in Big Tree Farms in Indonesia and Amma Chocolate in Brazil, is already there, too. He bought forgotten land, converted it to organic, and planted a plantation of 150 hectares of old original Java Criollo trees. He expects to see the first fruit in 2014.

Brett Beach could do the same in Madagascar to propagate that pure Ancient Criollo. But to him and many others elevating purity isn't, and in most cases can't be, the goal of genetic identification, nor should it be. For Brett, the importance for the future is not whether a field is 100 percent Ancient Criollo. Instead, he imagines fields with many flavor tendencies—fields where scientists help create trees and farms of 70 percent Criollo and, say, 30 percent Trinitario that are compatible and cross-pollinate, and that if managed right will at harvest have what Brett calls a rainbow of flavors. "That's part of what makes Madagascar so good: rainbows of flavor that give it the depth. I'm not saying don't do the hundred percent, but maybe try to control all the varieties you're pulling from, to create diversity and consistency of flavor. If that's what you want, then that's a good thing, too."

No matter what the clusters are named or how many there are, Gary Guittard is simply delighted there is still, for now, a rainbow of flavors to talk about. His family business was there in the mid-twentieth century when fine flavor was in decline but still more than 10 percent of the cacao grown (down from around 50 percent at the turn of the twentieth century). Over the next forty to fifty years, the company watched flavor become even less important, as chocolate grew more commercialized around the world and manufacturers used more of the bulk cocoas. And when it became very competitive economically and prices dropped, no one—manufacturers or consumers—demanded the expensive fine flavors, and people started growing ordinary varieties or just stopped growing cacao.

"In the United States," Gary says, "It wasn't until Scharffenberger in the 1990s that you had a reinvigoration of people realizing, 'Wow, there's actually different kinds of flavor in chocolate. This chocolate tastes different from the chocolate I've tasted! What's going on here and why is this different?' Today, with people so interested in where their food comes from and how it is grown, and heirloom varieties in crops like tomatoes, we have renewed focus on food and flavor and a new opportunity to map flavor starting at the genetic level. We have lost so much since bulk cocoas replaced flavor on the map. Now I think the public is ready for the focus on flavor on all levels."

A focus on flavor starts at the genomic level, in that Ancient Criollo leaf from Madagascar, packed in a plastic bag of white powder and shipped to a USDA lab in Maryland for identification. That DNA has the potential to help shape the flavor of chocolate in the future, from cultivation to consumption. "That's really neat," says Brett Beach. "That's closer to making it a reality for all of us. I think it takes time for it to sink in. Having that shared audience and having people passionate about it makes it real because no matter what you do, if you do it alone it doesn't really have much meaning. I've started to realize what this might mean. It's a big part of telling the story of what is going on inside forests."

In those forests as well as fields, plantations, and jungles, organizations worldwide are using genetic identification to build better breeding programs and give growers improved access to quality planting materials. But to ensure that fine flavor is a priority in all growing regions in the future, new and established flavor trees have two more immediate problems to deal with: the diseases that prey on them and the high-yield clones and hybrids that threaten their propagation. In fact, in the case of one high-yield clone, the two problems are even connected.

The Future and CCN-51

The release of the CGD was never about genetically engineering cacao, but that does not mean it is not happening. There are probably people working in labs worldwide, trying to "build" cacaos that—like industrial corn genetically modified organisms (GMOs)—will serve no higher purpose or concern than prolific production. For example, C-spot.com (a premium chocolate information and review site), in a 2011 post titled "Designer or Disaster Chocolate," was one of several sources to report details from Peter Schieberle's presentation at the American Chemical

Society (ACS) conference on the science behind the aroma and flavor of cacao. A professor at the Institute for Food Chemistry at the Technical University in Munich, Schieberle's research showed that despite the hundreds of flavor compounds in cacao, only a dozen or so were needed to simulate real chocolate flavor for taste testers—a potential path to manufacturing real-tasting artificial chocolate in the "cozy confines of a climate-controlled lab." Now, in all fairness, C-spot also reported that Schieberle offered tools for artisan manufacturers to understand how flavor works for improved and targeted flavor processing. But the findings sounded more like better cacao through chemistry than the kitchen science of foodie favorite and "Curious Cook" Harold McGee, whose work on the chemistry of food and cooking often focuses on flavor.

While the discussions about the dangers of GMOs in food are as complicated as they are polarizing, no one denies that GMOs are real, and it is impossible to prevent scientists from achieving genetically engineered results on cacao in the future. The reality is that ingredients like genetically modified sugars already exist. In response, Theo Chocolate became the first of what is likely to be many manufacturers who find it necessary to proactively address customer concerns about GMOs in chocolate. The company announced in the fall of 2011 that as part of its overall efforts to be as transparent as possible, all of its dark chocolate had been certified by the Non-GMO Project, a nonprofit collaboration of manufacturers, retailers, distributors, farmers, seed companies, and consumers dedicated to preserving and propagating natural and organic product choices. (We'll discuss the broader significance of this and other certifications in Part Two.)

Despite concerns, GMOs are not immediate threats to the future of fine flavor cacao or cocoa and may never do anything more than fund new science facilities at Big Candy. The more immediate threats and cautionary tales are the high-yield clones and hybrids that have replaced and are replacing fine flavor cacaos in their largest growing regions. Now, for the sake of those who still reflexively cringe at words like hybrid and clone, remember what Michael Pollan taught many of us who are not in the food industry: hybridization and cloning are not synonymous with GMOs, which have altered genetics through things like gene splicing. Heirloom fruits and vegetables are by their very nature hybrids. And cacao cannot seed itself genetically—remember those promiscuous cacao flowers? Grafting, cloning, and crossbreeding can reproduce a cacao tree's production, disease resistance, *and* bean flavor properties. But breeders often ignore or disregard flavor.

"I think there were and are a lot of folks who purport to develop hybrids with regard to flavor," says Gary Guittard. "You will have people who will say 'Flavor, flavor, flavor, flavor, flavor' but the reality in the fields is flavor has not been that important an aspect for decades. Production has been. Ghana is a great example of what happened in the 1970s when they went and replanted most of the country and lost that West African flavor— even more so in the last few years because they have integrated more hybrid trees. They have lost most of the older trees. If you look at the flavor of Côte d'Ivoire versus Ghana they are not the same anymore. In fact, in some places Côte d'Ivoire is considered a flavor bean."

For many in the industry, the most conspicuous example of the kind of high-yielding, hardy clone Guittard alludes to is

CCN-51 in Ecuador and Peru (though Brazil, Africa, and other countries also grow the clone on a smaller scale). Simply put, CCN-51 is a producer. It can produce anywhere since it does not need to be shade-grown in its early years like most flavor cacao and is tolerant to disease and difficult climate conditions. To date, it has the highest sustained production record of any cacao anyone has ever planted anywhere, outperforming all but the more recently planted and far less widespread Super Cacao in Ecuador by three to five and by some reports by a factor of ten or more per harvest, which for CCN-51 is twice a year. Average production in Ecuador is more than 2,000 kilograms per hectare (dry weight, fresh beans weigh about three times as much as dried) and losses to disease are much smaller than any flavor bean. In addition to its fertility, it also produces great rootstock for grafting.

With results like this, it is not surprising that CCN-51 is grown in addition to or instead of flavor. On the face of it, there is no reason a farmer wouldn't want to do this. There are problems, of course. CCN-51 requires more labor and maintenance, not to mention lots of water, chemicals, and fertilizer because its root system rapidly depletes the soil of nutrients. CCN-51 also takes up to seven days to ferment as opposed to as little as two or three days for Ecuador's flavor beans, Arriba and Nacional. Yet despite the increased time, labor, and maintenance costs, and even given the much lower price differential of CCN-51 on the cacao market, the income is higher than most farmers have been earning with the flavor beans that CCN-51 is rapidly replacing.

Another nail in the fine flavor coffin, another step toward a genotypic flavor map that is all gray? In fact, many people in the chocolate industry compare a future of CCN-51 in cacao to the evolution of another food that has resulted in a less flavorful world: the supermarket tomato. As exposed in Barry Estabrook's Tomatoland: How Modern Industrial Agriculture Destroyed Our Most Alluring Fruit (2011), the rise of the perfectly red and round year-round supermarket tomato in the United States was the result of the success of industrial varieties that have more than tripled the yield of the plants but at high environmental (fields are sprayed with pesticides and worse), human (almost slave-like working conditions to reduce labor costs), and flavor/nutritional costs. Genetic or crop variation and labor conditions be damned: let's grow fields and fields of this one tasteless version and nothing else. Give the people what they want!

The comparison of the tomato to fine flavor cacao is not perfect by any means. The human costs in Tomatoland are more comparable to those recently exposed among bulk cacao producers in Africa. But cacao is not an industrial crop harvested by migrant workers—the biggest farms, cooperatives, and plantations may be corporate-owned but are usually maintained by individual farmers. And CCN-51 was not developed in a lab; it was grown on an estate. The CCN in CCN-51 stands for "Colección Castro Naranjal." It is named for the Ecuadorian cacao breeder Homero Castro, and "51" is simply the number of the Trinitario-Nacional hybrid (a three-way cross) that had the greatest success of all he created. (Castro created the hybrid in the 1960s but died before he could patent it, which is one of the reasons CCN-51 is so widely available.)

With today's access to technology and research increasing our capacity to receive, analyze, and understand data and chart more productive paths with less trial and error, more powerful

CCN-51-like clones could fill the majority of the fields in Ecuador instead of just the nearly 60 percent it does now. (In May 2012, Bloomberg News reported that Armajaro Trading Group Ltd. of London was "investing $1 million over three years to increase production of fine cocoa beans in Ecuador," but according to Armajaro, CCN-51 accounted for "110,000 metric tons of the 190,000 tons of cocoa the country will produce" in 2012.) If so, CCN-51 will likely turn out to be only the current step in the evolution of these clones. Enter Super Cacao. As Jeffrey Stern reports from Ecuador, "One program run by two Americans has identified Nacional trees that perennially are super producers for whatever reason, and are now propagating these trees. They are calling it 'Super Cacao' and its first hectares show it can far exceed the CCN-51 but may also have some fine flavor."

So it might be fair to ask at this point: How does CCN-51 taste in chocolate? Most people in the fine flavor business we spoke to called CCN-51's flavor anything from a polite "not fine flavor" to "horrible," "crap," "a disaster," and "acid dirt." C-spot calls it "weak basal cocoa with thin fruit overlay; astringent & acidic pulp; quite bitter beans & generally sub-par quality." But that hardly puts CCN-51 on par with that awful tomato, because things are not so black-and-white in cacao. CCN-51 has more than just a passing resemblance to its fine flavor cousins: It has a flavor heritage and is grown in regions that are traditionally associated with flavor. While it will never be a fine flavor bean— though it was sadly passed off to farmers as such—the reality is a number of companies are looking at alternative fermentations of CCN-51 for the future and some sectors of the industry tout it as a flavor clone. And there's the real difference between CCN-51 and the Tomatoland tomato: genetics is only one piece of any cacao flavor puzzle. Once it is harvested, the tomato's process is done. Even Jeffrey admits, "If CCN-51 is properly fermented, and fermented separately from other varieties, and then blended with some Nacional beans that have also been properly fermented? You could produce something pretty decent."

The tomato story certainly offers a warning and, interestingly, an idea of how we might preserve a tastier future. The rise of farmers' markets nationwide in the United States, an heirloom and flavor movement among growers, increased awareness of GMOs as well as growing public concern about origin, seasonal eating, and how our food is produced, have helped restore some balance in the tomato world—at least at the high end of the market. To paraphrase Marion Nestle, people are voting with their forks and building awareness of and resistance to the supermarket tomato.

This possibility exists in the cacao fields, too. Genotypic flavor identification can ensure that beans are specifically and honestly identified and can add value through flavor differentiation. In other words, there will be less chance to pass CCN-51 off as something it is not. The fine flavor industry is working with farmers to get them better quality materials for planting, to increase yields on their existing fine flavor trees, and to help them understand the value of their fine flavor beans. As a result, farmers are regaining national pride in the history and origin of those traditional flavors, and using the stories behind their products to better market their quality. And with the fine flavor market expanding just as quickly as the rest of the chocolate market, and those fine flavor beans commanding more and more of a premium, Ecuador is seeing some shift away from

CCN-51. Not to a better CCN-51, but toward Nacional and other populations (as is Peru).

"You go through Ecuador today and there are signs with pictures of a pod that looks like CCN-51 with a circle and a slash through it," reports Joe Whinney from one of his recent visits. "The entire nation once mobilized around CCN-51 and bulk buyers will keep buying it, but now some farmers are starting to differentiate. Farmers, if given a choice, will usually want to choose a path of national pride, if they can survive. So if there is a development opportunity because there is a real market, I think we can all work together to move the needle in a few years." Many in the fine flavor industry feel the same way, and strive to improve the lives of the farmers they work with as an essential part of preserving and propagating flavor that has as much to do with genetics as it does doing well by doing good—for the environment and especially the farmers. Only then, many feel, will farmers turn away from planting CCN-51 and toward Arriba/Nacional trees—but only if there is productivity from the trees, a market for the product, and a premium paid for the flavor genetics. Big and small seem to agree that there is. Even Nestlé stepped into the fray in 2009 with its "Cocoa Plan" and is now working with farmers to replant Arriba/Nacional trees with genetics that hopefully will have productivity to rival CCN-51 in the future.

"The reality is these new clones are coming," says Santiago Peralta, cofounder of the Ecuadorian organic manufacturer Pacari Chocolate. "What we're destroying is that genetic bank of cacao. We have every cocoa tree from the old cacao in Ecuador. What we are only seeing is the short term and not the long term. Long term what we need to have in nature is a variety. If there is ever a disease that affects these clones, then the result is going to be a catastrophe because it is all monoculture. So what we are trying to do is restore a balance and have trees that have a nice flavor and are local. And instead of putting only one tree, put a variety of trees—even some of these productive clones. Maybe every farmer would have one, two, five, even ten flavor trees, and they are local genetics, and then you have productivity and diversity. That for me is the key. It is not just a matter of productivity in the future but awareness of the importance of diversity."

But before we can get to that future, Santiago mentions the other big problem that must be dealt with in order to stem the tide of CCN-51 and make it possible for fine flavor to be sustainable from an ecological and economic perspective. The reason CCN-51 could be planted all over Ecuador in the first place: all cacaos, but especially fine flavor cacaos, are susceptible to disease.

There's a Fungus Among Us

Dr. Wilbert Phillips-Mora is a renowned expert on cacao diseases and breeding at the Genetic Enhancement Program of Cacao at CATIE (the Tropical Agricultural Research and Higher Education Center). His warm disposition and welcoming personality never fail to charm the chocolatiers whom Ecole Chocolat brings to Costa Rica every year, which is good because the stories he tells about the diseases affecting cacao would be awfully scary if told by someone who sounded as intimidating as, say, James Earl Jones. The descriptions are ghastly: frosty pod rot gets its name from the way the cacao pods look after being infected (like they are covered in a layer of frost in the middle of the tropics), and witches' broom is named for the way the tree looks after the disease has attacked (like a gnarled broom sticking out of the earth). Both are fungal pathogens that spread through airborne spores. Their scientific names alone—*Moniliophthora roreri* and *Moniliophthora perniciosa* (formerly *Crinipellis perniciosa*)—could give someone nightmares if spoken by the man who voiced Darth Vader.

Truth is, despite Wilbert's gentle delivery, those nightmares are well founded. Coupled with the smaller problems of insect and environmental damage, these cacao diseases have been nightmares for nearly a century and currently account for more than $500 million in cacao losses every year. And the fine flavor regions of Central and South America have been especially hard hit.

It was a one-two punch of frosty pod rot (often called monilia) and witches' broom that hit Ecuador in the early nineteenth century and wiped out Pure Nacional, then the world's dominant and most sought-after fine flavor cacao. That's why Pure Nacional was thought to be extinct until Dan Pearson found it in Peru, and why CCN-51 succeeded it, rather than replaced it, in Ecuador. While frosty pod rot had first been recorded more than a century before in Colombia, until it hit Ecuador the world had never seen devastation on that scale. Nothing had been so thorough in its destruction, save perhaps the unknown natural disaster (some say massive hurricane) in 1727 that took out nearly all the Criollo in Trinidad and Tobago and most of the cacao in the Caribbean. But at least that disaster had a silver lining for flavor: the introduction of Venezuelan Forastero to the island, which when crossbred with the remaining Criollo trees led to the creation of the Trinitario hybrid. Few would call the arrival of CCN-51 decades after the devastation wrought by frosty pod rot a silver lining unless you consider the alternative of no cacao at all.

While neither frosty pod rot nor witches' broom kills its victims, yield losses are so extreme that what's left is not worth maintaining and farmers often abandon the industry, which was the case with any remaining Nacional in Ecuador by 1925 and then again when frosty pod rot spread north to Panama. Following the Panamanian outbreak in 1956, Wilbert recounts in *Fifty Years of Frosty Pod Rot in Central America,* the disease spread to Costa Rica in 1978 and then rapidly through eight more countries before its arrival in Mexico in 2005. According to Wilbert, frosty pod rot has "limited geographic range [but] unlimited potential for damage," and without aggressive management, even the most disease-resistant trees can lose up to 30 percent of their pods. Less resistant varieties, especially those planted close together in fields with little diversity, can lose as much as 90 percent of their yield, which was the case in Costa Rica. Frosty pod rot turned the country from a net exporter of cacao to a net importer in less than a year.

Steve De Vries saw what unchecked devastation from frosty pod looks like in Mexico when he led a trip there in 2011. The disease may be hitting the country a little more slowly than it did Costa Rica, but it still has resulted in production drops from 50,000 tons a year to around 20,000. Couple these losses with the country's high labor costs, which in the Mexican state of Tabasco is the result of competition with the petroleum industry, and even the farmers who survived are turning away from cacao. On the bright side, this could be a potential boon to the flavor industry in Mexico as those farmers who remain in the cacao business are turning to flavor beans to get the price they need to survive. But as De Vries says, "Their success depends on how the monilia shakes out. We saw several places that wanted to do this but looked like they would never make it because the monilia was just all over the place. So now some of those places are just pulling out the cacao and putting in pastures for cattle."

Witches' broom is also in Mexico but affects primarily lime trees at this point. It has mostly stayed south of the Panama Canal in cacao to wreak havoc in South America, causing losses of up to 75 percent in infected trees. No story of its damage is more gruesome than what happened in Brazil. And industry expert Chloé Doutre-Roussel proved in her presentation at the January 2012 FCIA conference that not even the allure of a French accent makes any of the following details more pleasant to hear. Brazil is just now recovering from a twelve-year battle (1985 to 1997) with witches' broom, which resulted in 70 percent cacao loss and the country, like Costa Rica before it, going from net exporter to net importer.

Even more troubling is the fact that the witches' broom attack in Brazil is now commonly regarded as an act of bioterrorism. Reported by Chloé in the October 2011 article "Brésil & Balai de sorcière: un crime, pas un accident?" ("Brazil and Witches' Broom: A Crime, Not an Accident?") and first brought to light in an article from *Veja* magazine, six people connected to the Workers' Party attacked several plantations in southern Bahia with witches' broom in order to undermine the political influence of the "barons of cocoa" and destroy their livelihoods. The plot did not succeed in claiming power but they did destroy one of the more vibrant cacao industries in the world. Replanting was unsuccessful. Land was abandoned. Suicides became common. And a major agricultural industry in the world's fifth largest country was brought to its knees.

Terrifying stuff and a very, very real threat now and in the future—so real that the Dominican Republic, one of the rare countries where no cacao disease is present, closed its borders to all cacao material and any genetic and agricultural research. "This is a very privileged condition that obviously people want to keep and maintain," says Massimiliano Wax of Rizek Cacao. "We are lucky right now: we have a good mix of genetic material, but we cannot say that our cocoa genetically is as good as some regions in Venezuela or regions of Ecuador. Thanks to good selection and good investment in postharvest, and also education and training and certification to the farmers, we are able to score some flavor points. But there's still a lot to do in terms of genetic research to improve these breeds. This is really a problem for the future. We should be able to find a safe way of researching and bringing in genetic material. But then researchers look at Costa Rica and the monilia, and legislators think of Brazil and see this terrorism is very easy to do, so they refuse. We are stuck no matter the strong desire."

Overreaction? Not according to Dr. Lyndel Meinhardt of the USDA-ARS, an expert on plant pathogens. He says closing borders is "probably the only way right now" to absolutely deal with these fungi: "You never know what might happen with the movement of these pathogens. Someone picks up a pod and without even thinking transports it to another place and spreads the disease. That's what we saw with frosty pod where the pods went to another country, and thinking it is not going to be a problem, and all of a sudden it starts to rot and they throw it away and it produces spores, and the next thing you know you've got airborne spores and an outbreak of frosty pod. And you can't control it. You can only manage it, and manage it in the right way for your particular environment and for the particular pathogens that you're facing. For the immediate future that's how you can still get the yields. It's got to be a managed crop."

The problem is that cacao is not traditionally a managed crop. It is historically an amalgam of orphan crop, neglected crop, and underutilized crop. This is why even the most optimistic people in the business believe it is not a question of whether one of these diseases will jump to another country or continent in the future; it's a question of when and where.

Southeast Asia has some growing problems, and the insect known as the "cocoa pod borer" causes considerable damage for a season (growers in Papua New Guinea recently had flavor crops destroyed by it), but losses are not comparable to those from frosty pod rot and witches' broom; they do not exist there . . . yet. But the environmental conditions are similar to South and Central America and frosty pod rot or witches' broom could thrive if introduced. Right now, Western Africa has some major labor and fungal problems (brown pod rot), but none caused by these major pathogens. Moreover, the drier air and soil conditions in Western Africa are not necessarily conducive to fungi development. Nevertheless, Robert Peck of the World Cocoa Foundation (WCF) notes that some governments in parts of Africa, notably Ghana and Côte d'Ivoire, do not allow any genetic material in for the same reasons as the Dominican Republic. What *if* frosty pod rot arrived in Western Africa? Dr. Wilbert Phillips-Mora is not alone in believing the cocoa industry would be unprepared and face collapse.

Wilbert tried to address this lack of preparation by requesting funding for a crew in Mexico to study how frosty pod rot behaves in a new country. He wanted to try to predict its behavior and develop strategies to deal with it if it spreads again. But he was rebuffed—a missed opportunity that can only be remedied in the future by an outbreak somewhere. Until then, many people like Wilbert are on the ground in South and Central America—from government-backed groups to international corporations and aid groups to individual manufacturers—to manage the current situation. Scientists like Lyndel are on the ground, too, studying the interaction between the pathogens and the trees to get a better understanding of what actually constitutes and causes the disease, to help that management and see if there's anything that can be done genetically to alleviate or moderate disease interaction in the future. And farmer education is also playing its part, from basic management techniques to introducing genetic and biodiversity (i.e., planting a variety of cacao and other crop materials because diseases thrive in homogenous fields).

Dr. Darin Sukha at the Cocoa Research Unit at the University of the West Indies, who has worked on projects in his native Trinidad and Tobago as well as in Belize, Grenada, Jamaica, Dominica, and West Africa, sees this return to diversity as extremely important for sustainability of cacao as a whole and fine flavor specifically: "The loss of genetic diversity in farmer selections in the 1960s and 1970s was disheartening, but now we have an awareness of the diversity that exists. There is a desire to capture this old material in ancient growing regions as a sort of linchpin of the genetic diversity, and use that as the basis for fuller breeding." That's why the Cocoa Research Unit leads expeditions into the Upper Amazon to find varieties in the wild and test them for disease resistance and flavor. Says Darin, "These traditional varieties may not be the most prolific but by virtue of their being around for so long they would have surely displayed some level of adaptability to disease conditions as well as the environments that prevail in the countries or areas that they are found."

Dr. Wilbert Phillips-Mora agrees: "In Central America, Ancient Criollo is the high-quality material people talk about, but it is among the most susceptible to disease and could actually spread the disease. The material is interesting in terms of quality but the risks are great to the farmer. The basic data says that 70 percent of the cacao produced belongs to just one of the ten genetic groups of cacao—70 percent is Forastero! Using the wide genetic diversity of cacao could solve most of the problems by introducing genes that fight against monilia and incorporating them and seeing how the cacao develops. In other words, creating a blend at the genetic level, no different from what a chocolate maker does in manufacturing to get the flavor profile they want: combine the traits of production, disease resistance, and quality through breeding and produce a kind of polyclone."

But progress in breeding these polyclones has been slow. Interesting opportunities to work with the genetic diversity of cacao have been lost. Some of the slowness was due to a lack of funding from governments and other organizations after decades of depressed cocoa prices in the 1980s and 1990s that followed the extremely elevated prices of the 1960s and 1970s. Some of it was the result of little collaboration and shared database material between collections (detailed by Gary Guittard and others at the start of this section). And some of it was because of poor access to genetic information at the molecular level until the Cacao Genome Database (CGD) made the sequenced genotype available. Progress was even slow at CATIE, which holds a special place in the CGD story: the sequenced genotype in the CGD is Matina 1-6, a Forastero clone from the CATIE collection that used to thrive on Costa Rica's Atlantic coast until frosty pod rot destroyed the population.

The main reason progress in breeding has been slow, however, is that genetic breeding and improvement in perennials such as cacao *is* a painfully slow process. Producing one variety for farmers could take up to twenty *years*. It took CATIE fourteen years to establish its clonal gardens in Central America. "The result of all this slowness is there has been a complete lack of variety for breeding programs and farmers," says Wilbert. "There are a few new varieties available worldwide—some from Papua New Guinea, some from Ecuador, some from CATIE—but they're not very common. If you check the material being planted at this moment in the world, we are using exactly the same material that was recommended forty years ago by the original researchers."

So how do we improve this breeding to preserve and propagate flavor?

Some in the industry think we don't have to—that supply is meeting demand and that enough is being grown for the future. Many manufacturers believe that to continue to grow and meet consumer demand for higher-quality chocolate, they cannot expect to get their fine flavor cacao from the countries affected by disease problems. Now, no one we spoke to believes that any genetic advancements will allow cacao to be grown sustainably outside the 20/20 belt or zone—the area twenty degrees north and south of the equator where cacao grows worldwide—any time soon, if at all. But internationalization is key. Flavors must move to other parts of the world where there is more opportunity—more farmers with the incentive to grow fine flavor cacao and fewer disease problems. For example, Hawaii and Australia flavor cacao breeding programs have started to yield promising results. Joe Whinney of Theo Chocolate is looking at long-term development for flavor growth in Africa, specifically Tanzania and the Democratic Republic of the Congo, which are less developed from a fine flavor standpoint: "We are putting together grafting programs that can allow us to produce some cacao of the highest quality, and there is real incentive for farmers there. Growth opportunities are not in established countries. In the scheme of things, I think in three to five years, you can have a huge impact on one region for the small part of the industry that is fine flavor."

Dr. Pham Hong Duc Phuoc, director of the Biotechnology Center at Nong Lam University in Vietnam agrees. Vietnam, which has no major disease problems (yet), had no cacao tradition until Phuoc started working on the crop in 1997. He started with some small grafting programs to clone the cacao and increase the quantity. When those programs succeeded, the WCF sponsored larger programs, which provided him with the opportunity to expand all over the country and test cacao's adaptability to various environments. Today, with the industry established, maintaining plant diversification has become one of the highest concerns in the region. As a result, Phuoc sees widespread benefits from genetic testing to help maintain this diversity. He currently has germplasm of Trinitario, Forastero, and a few Amelonado clones and is experimenting with white

Trinitario and other fine flavor clones, especially in the Ben Tre province.

The potential of these flavor programs and the collaborative spirit that now pervades the industry is exactly why people like Steve De Vries are not concerned about losing flavor production in the face of these disease problems and breeding issues. "Listen," he says, "if you look at the history of cacao, whole countries lose all their production for twenty years but another country comes up. There is always cacao. The important part is that everybody thinks of being in this together. That doesn't mean everybody has to show his or her undershorts, but the people who share with other people and help build the community will be successful and the ones that don't, won't. The more you know and can help people, the more they can help you."

Meanwhile, there are scientists in the labs looking for solutions in the sequenced genome. For example, professor Mark Guiltinan at Penn State University is studying how genome sequencing accelerates the breeding of disease-resistant plants. As reported in the fall 2011 *NCA* [National Confectioners Association] *Journal* article "Genomic Breakthrough Speeds Cacao's Evolution," Guiltinan and others at Penn State have been researching the quantitative trait loci responsible for things like flavor and immunity in cacao and that have the most influence on the bean. Their work resulted in the discovery of NPR1, a "master regulatory gene" that controls the entire cacao immune system—a huge step in perhaps releasing disease-resistant varieties to the public in the future.

But in South and Central America, the industry needs help right now. There are few places to start anew and escape frosty pod rot and witches' broom. Breeding new genetic material in the future is essential to preserving and propagating flavor—and providing a higher standard of living for the farmers who grow fine flavor cacao. But farmers cannot wait twenty years for another round of "this seedling should work and might even taste okay." Thanks to the genetic research of the last few years and the growing collaborative spirit in the industry they may not have to.

From Farm to Lab and Back:
Breeding a New Future for Flavor Through Genetics

Historically, cacao breeding follows this course: Breeders take the pollen from a father tree and pollinate by hand the flowers of, say, 100 mother trees. The next harvest, they take forty seeds from one pod from each mother tree and create seedlings for 4,000 trees. Now the breeders must wait five years before the trees bear fruit. Only then can they determine whether the parent trees transferred the desired traits to the next generation. They usually find that *three or four* trees inherited all the traits they wanted. That's a seven-year cycle just to get productive seedlings from those three or four trees in the soil. And it will take up to another seven years to get the fruit.

While genomic methods and research cannot produce cacao trees that yield fruit in fewer than the typical five to seven years or make the fruit less labor-intensive to harvest, they should be able to help quickly and efficiently breed new trees. Modern genetic tools—specifically marker-driven selection—can help cacao breeders be much more productive, and potentially transform the industry. The tools will not replace the hand pollination, harvesting of the seeds, and planting of 4,000 little bags of soil. But today breeders do not have to wait five years to test the fruit, just two weeks for the seedling to grow some fresh leaves. Then, in a process no different from Brett Beach's sampling of the mature leaves of that Ancient Criollo in Madagascar, breeders clip pieces from the seedlings' leaves, tag them, and send them to a lab to extract the DNA. From there, it is an easy process (for scientists, that is) to compare them against the markers or traits the breeders are looking for from the mother or father. And just as important: they can avoid those traits they do *not* want. Thus, when the seedlings are a month old, breeders will know which three or four of them have all the traits they want. They can immediately grow those out in a greenhouse while they head out to the field and do another cross, because they now have the tools to do a cycle in *every* crop year. That's the power of marker-driven selection.

In the fall 2011 *NCA Journal* article, Ed Seguine detailed some of the progress being made at the genetic level and argued that marker-driven selection is an essential step to help cacao remain a desirable business for the growers, with more productive and disease-resistant trees. Ed anticipates that in fifteen years, modern genomic methods will produce "ultra plants with significantly changed yield potential and disease resistance . . . with many more plants at the upper end of yield and disease-resistance potential because of the tools from the genome publishing." As a result, a four-hectare farm that produced 400 kilograms of cacao on each hectare could produce 4,000 kilograms on just one hectare with noble woods and food crops grown on the others, resulting in a better income for the farmer: "This will dramatically change the economic stability of cocoa farming and create an agricultural business enterprise that looks at the short-, medium-, and long term."

But how will it taste? Signs are promising. A cup of *chocolat chaud* might be in order. Flavor has started to play a more prominent role in cacao breeding—not to make over the bulk cacao fields of Côte d'Ivoire and Ghana or to create better-tasting super-sized CCN-51s, but to grow sustainable cacao with the potential to make great chocolate.

Prior to this century, a lot of breeding programs focused on "mixing"—trying to bring in markers from as many different cacao types as possible and "pyramid" or stack multiple resistant or trait genes on top of one another. The resulting high-yielding, disease-resistant lines often lost their great flavor traits. In other words, the process bred out the flavor. "When I started working in cocoa, flavor was almost an afterthought in plant breeding programs," says Dr. Darin Sukha. "You would breed for high-yielding varieties. You breed in disease resistance and good tree habits and then at the end of it all there would be almost a cursory attempt to say, 'Oh, now let's make sure it tastes okay.' Over the last ten years, heightened awareness of flavor as the final expression of cocoa quality has placed flavor testing at the start of the selection process. I am very heartened by this focus on flavor and quality in many breeding programs."

We have already mentioned some examples of this focus: Dan Pearson propagating Pure Nacional in Peru, Frederick Schilling planting Java Criollo in Java, Joe Whinney grafting in Africa, and Dr. Pham Hong Duc Phuoc's most recent work in Vietnam. Another example is the USDA-ARS, which has started evaluating its breeding programs with fine flavor in mind. According to Dr. Lyndel Meinhardt, this involves looking beyond stacking genes from different types of cacao in the future: "Maybe we should go back to the original population and the original areas where these fine flavors are and stay within a set population to try to find those traits within that population, so that when we do make those crosses, we are not bringing in a lot more differences but maintaining some of those traits. The point is if you're doing the breeding, I think flavor has to be one of the main traits that we're constantly looking at, rather than simply saying, 'I've got a great producer that's going to yield 3,000 kilograms per hectare.'" But, as Gary Guittard cautions, we have to be careful about our biases and open to all possibilities: "We can't just say, 'Forasteros are not flavor.' There are some great Forasteros. I don't think people realize that the Forasteros have as much individuality, flavor, and history as the Trinitario and Criollo, and could be the foundation of fine flavor in the areas most affected by disease."

At CATIE, Dr. Wilbert Phillips-Mora started to work on clones with disease-resistant, high-yield, and good-quality components in the late 1990s. Until then, CATIE's breeding strategy focused on the same qualities most programs did: to identify the sources of resistant beans and to create new material using the traditional breeding strategies—that painfully slow trial-and-error process that can last a generation or more. To save time, Wilbert, who is in charge of the breeding program, "jumped some steps" by making educated decisions based on the perennial data and adjusting quickly to any perceived changes. After around twelve years, once CATIE was successful in identifying and creating a new genotype that had all the desirable markers and traits for superior disease resistance and high yield, it started its research on quality. But even with the latest genetic tools today, producing these crosses or polyclones is a very difficult balancing act, just as it was a decade ago when CATIE started in this direction.

"The world had changed in terms of cacao," says Wilbert. "Opportunities in our industry for diversity and identification of product for determination of quality traits like high polyphenol content or high fat content were open for exploration. The possibilities for breeding were huge. But the demands for quality were recent. We had that exact same material as forty years ago in most countries. There was no diversity. We had no genome sequenced, poor facilities for the kind of research and analysis required, and poor maintenance of data in the cacao collections worldwide if it was even shared. It was not possible to produce that material to meet the quality demands quickly."

To transcend these limitations, Wilbert drew on the collaborative spirit and got in touch with different companies around the world that could help carry out the analysis needed and speed things up. These relationships provided CATIE with the opportunity not only to carry out a traditional breeding program, but also to use the latest DNA technology to make rapid progress. CATIE got support from, among others, the World Cocoa Foundation for access to and development of diverse materials and the USDA for planting and collecting information in the field, which would then be sent back to the USDA for molecular analysis. And one of the first companies that cooperated with CATIE from the start was the Guittard Chocolate Company, which signed on to do qualitative evaluations of each bean's flavor potential—a pioneering connection that is now being emulated by Mars and others. "We have to be willing to listen," says Gary Guittard. "We must try to understand the hybrid area so that we can learn from them and they can learn from us."

Theo Chocolate, Felchlin, and others later joined Guittard in analyzing the six CATIE clones (all of which are Trinitario type) with the most potential for disease resistance, yield, and quality. After the beans were properly fermented and dried by CATIE, the chocolate manufacturers evaluated them for fat content, aroma, flavor, and color. Guittard found all the beans had good fat content, and all the manufacturers agreed in their evaluations of the flavor profiles, with one of the six thought to be of low quality. Two of the clones (PMCT 58 and CATIE R1) were deemed high quality, and two of them (CATIE R4 and R6), which had much higher disease-resistant traits, were found to be very interesting—so interesting that they were selected as two of the ten most important clones at the Salon du Chocolat in Paris. And in the most important news for the future, all are now established together in small clonal gardens in six Central American countries: Panama, Costa Rica, Nicaragua, Honduras, Guatemala, and Belize. CATIE has partnered with the Cacao Centro América Project to plant the gardens and evaluate how the clones perform in different environments so they might become the source of material for new plantations. Thanks to the speed now available to evaluate new genetic material, local field trials are also underway in each of those six clonal gardens on thirty new clones eyed for future plantations, many with strong flavor qualities.

Why six clonal gardens across Central America and not just one plantation in Costa Rica? Because fine flavor cacao reacts differently every place it is planted. A tree that grows nicely in the Dominican Republic will have different challenges in other regions. You can't just take cacao trees and move them someplace with a different altitude or ecology. That's a path to neither flavor nor ecological sustainability. According to Wilbert, this is the best way to ensure that the cacao will be perennially stable when the material is released to the public: replicate the conditions in which the cacao will be grown.

And this is the ultimate goal: sustainability for the farmer through cacao that can resist disease, generate high yields, and command a premium price for its flavor. "It is not possible to wait," says Wilbert. "Our farmers urgently need these varieties." And in turn, scientists like Wilbert and the industry as a whole need the farmers: they are the key to all of our success. In other words, genetics means nothing for the future of fine flavor cacao if no one wants to grow it. Sometimes you have to wonder why anyone would. Cacao is an environmentally sensitive and labor-intensive crop that is susceptible to disease and insects in the *best* conditions. Half of all cacao production in Central America takes place in the most isolated rural areas and almost all of it is grown on small-scale farms of around two or three hectares and rarely more than five. Even the highest yields often don't result in a high quality of life for the farmers and their families. Truth is, fine flavor cacao needs to be ecologically *and* economically sustainable to survive—not only for the crop itself but also for the people who grow it.

That's why manufacturers—in word and, more and more, in deed—are making direct buying connections and relationships with the people growing their beans. They realize that paying more for fine flavor cacao is not enough if more money does not go directly to the farmers. And money alone is not enough. They are visiting the farms. They are educating farmers on the postharvest process. They are bringing the farmers chocolate made from their own beans. In this way, the market is less of an abstraction, but a network of relationships that leads to a higher quality of life for everyone. Dr. Lyndel Meinhardt, who comes from a farm family in Missouri and was a farmer for three years, understands this from the farmers' viewpoints: "We're not talking big industrialized guys that can go out and borrow a lot of money to line their fields and put on a lot of fertilizer like we do with crops in the [US] Midwest. These guys for the most part don't have a lot of money to invest in new inputs. I totally understand how the farmers look at this, and we as an industry have to be more connected to them, especially the fine flavor growers."

In the end, genetics solves only the first part of the fine flavor puzzle: how to make a potentially great tasting and hearty bean on a tree. Harvest and postharvest—particularly fermentation and drying—is where the work continues and the human touch is essential. Without love for the labor in the fields, chocolate manufacturers' labors of love to produce the world's best chocolate would, to quote *Casablanca*, not "amount to a hill of beans in this crazy world."

PART TWO

From the Ground Up: Farmers, Farming, and Flavor

Tell me and I'll forget;
show me and I may remember;
involve me and I'll understand.
—Chinese Proverb

Near-Death by Chocolate

Art Pollard woke up as his friend's car rocked back and forth, wildly snaking its way through hundreds of turns up the mountain's one-lane "highway." Rock wall and jungle lined one side of the road. There was no other side—just a drop of hundreds of feet.

The road narrowed as they climbed. Tight hairpin turns and blind curves seemed barely big enough for a single car to pass. And now it was raining. Torrentially. Rivers of water ran down the road and soon small landslides followed. The road doubled back on itself revealing a deep mountain chasm. Art looked to the bottom: a bus. It had clearly been there a while, but how long? How does a bus even make this trip? "That could have been us," Art thought.

Welcome to Venezuela's Henri Pittier National Park.

The descent on the seaward side of the park was longer, far less harrowing, and beautiful. As the rain diminished to a drizzle, the mountains and Caribbean Sea expanded before the windshield. Huge stands of bamboo planted years ago to keep the original dirt trail from washing away lined the road. Tiny shops appeared selling *arepas*, legendary Venezuelan corn cakes filled with meat and cheese. Art opted for tamales, which he ate as the car headed for the sleepy town of Choroni.

The eight-to-ten-foot swells that greeted Art in Choroni— easily big enough to tip the small boat hired for the thirty-minute ride to their final destination—didn't faze him after the mountain adventure. Besides, the only alternative was an eight-hour hike. Heck, he could swim, and nothing could dampen the excitement of being so close to Chuao, a little village on the Caribbean coast of Venezuela that Art calls the "home" of cacao—not because it is where cacao originated but because for four hundred years it has produced some of the world's finest and most highly sought after cocoa beans. Here, with historic precision, young and old work together to process the fruit, loading the beans into wheelbarrows at the fermentary, carrying them to the patio in front of the 200-year-old church to be dried, waiting until the beans are ready to be brought in, and then starting all over again. Art calls the crisscrossing between the church and the fermentary a "dance"—a "magical scene" performed with practiced skill. No wonder Chuao beans have a storied history and command premiums equal to or greater than any other and that, until recently, European companies had locked up exclusive rights to those beans.

And Art was heading there in November 2010 on a mission of utmost importance: to bring the people of Chuao bars of his Amano Chocolate made from their own legendary beans.

In a world where farmers at markets worldwide seem more and more connected to their products, Art's wild adventure may sound like the cocoa version of carrying coals to Newcastle. In reality, it is the exact opposite. Few of these skilled workers had ever had someone bring them chocolate produced from their labor. Not simply because of the remote location and the treacherous mountain road. Not only because those farmers prefer the sweet drinking chocolate enjoyed by most people in the Americas. Not just because fine chocolate is an expensive treat and the hot and sticky weather in Chuao guarantees a melting mess.

It's because the farmers are the proverbial low men and women on the chocolate totem pole—traditionally many, many steps removed from the end result for consumer, chocolatier, chef, and even manufacturer. We may celebrate fine chocolate bean to bar, but who's behind the bean? In the case of Chuao, it is farmers who are adept at production and fueled by generations upon generations of experience. They are masters of their work and excited by it. If full understanding and connection to flavor wasn't happening in Chuao as a rule, where *is* it happening?

Amano had waited only a few years for a chance to make chocolate from those incredibly rich, complex, and fruity Chuao beans. Most farmers, whether in Chuao or less renowned regions worldwide, have waited for generations to taste what chocolate manufacturers produce. Despite four hundred years of cacao history in Chuao, Art was doing what few men or manufacturers had done until recently anywhere in the world: closing the fine flavor chocolate circle of life by bringing the final product home.

The people of Chuao did not expect that chocolate from Art. But Art, like a growing number of fine flavor manufacturers today, wanted to bring it to them. "It just wasn't right that we had released this bar, but the farmers had not yet tasted it," he thought, and so he and the folks at Amano did everything to make the trip possible. That day in Chuao, he told the people that he was honored to work with their beans. He encouraged them to care for and treat them like their children—the chocolate made from them was *that* special. It brought joy to the palates and hearts of men and women worldwide. Then, to the delight of everyone, he produced the chocolate bars from his backpack and handed everyone a bar. The wrappers were decorated with a painting of cocoa beans drying in front of the church—their own "magical scene" brought back to them.

"If I had to go through the entire arduous journey to get to Chuao just to be able to spend only this brief moment with Chuao's farmers and see the pride in their eyes and their smiles, the whole trip would have been worthwhile," Art said of his experience.

This story will surely resonate with many but especially those in the fine flavor chocolate industry. Today, almost everyone in the world of fine flavor chocolate, and even many in the chocolate business in general, talk about how fidelity to farmers is essential to the industry's preservation. They may disagree on solutions and diverge in their business models and approaches, but they all agree on the importance of direct relationships, fair prices, social responsibility, and better processing at origin.

The definition of fine chocolate may be completely subjective—described by matters of taste as much as high-quality ingredients and the artistry of the manufacturer and chocolatier. But there is no dispute that cacao genetics, origin, and proper processing are the essential first steps to creating any fine chocolate. Start with a mediocre bean and you might create something good, but *fine*? No.

Having farmers taste the chocolate made from their beans is a critical component in helping them understand flavor and what is at stake. Or as Gary Guittard, CEO of the oldest family-owned and operated chocolate company in the United States, says, "Understanding flavor will add more fun!" In the end, it is no different for farmers than it is for the consumers who love the end product: to taste is to understand.

Tasting chocolate is thus a cornerstone of involvement and one way to add value and values to fine flavor chocolate; preserve, protect, and propagate flavor cacao for the future; and ensure that the beans are as good as they can possibly be. If this sounds like a simple starting point for a complex issue, it is. And nothing makes this clearer than a discussion of fair trade certifications.

The Power, Promise, and Realities of Fair Trade

For Jorge Redmond, CEO and president of Chocolates El Rey in Venezuela, the conversation with the German fair trade representative was getting confusing.

"What?" Jorge asked.

"How many toilets per square kilometer are there in your growing area?" said the representative.

"What are you talking about toilets? Have you ever *been* to a growing area? There is nothing there," Jorge said.

The man on the phone again stressed the need for toilets and Jorge tried to explain the centuries-old nature of cacao harvesting in Venezuela, its origins, and how he worked with his producers. Jorge wasn't sure the man understood this or any of the other points he made but at least this caller wasn't making accusations like the last one. "The first fair trade guy who called us up started by asking, 'Do you have slaves?' I got on the phone with him and said, 'Do you read history? The last time there were slaves anywhere in Latin America was two centuries ago. We are not Africa. I don't know what kind of stupid question you're asking.'"

Stupid questions aside, Jorge was on the phone that day because certifications of all kinds are proliferating in our food chain and show no sign of ebbing in the future. Sales of certified products, especially fair trade-certified organic, have increased by double digits every year in the twenty-first century. Most recent reports from the Fairtrade Foundation in the United Kingdom and Fair Trade USA reveal 30 percent growth of certified products in Europe with 48 percent of it certified organic, while North America has seen an 83 percent growth year over year, fully 90 percent of it organic. Not much of this is cocoa—yet; less than 1 percent of the cocoa worldwide is certified fair trade. But things can change quickly when big chocolate and candy companies get involved on a larger scale. When Nestlé certified the European Kit Kat fair trade in 2008, it more than doubled worldwide sales of fair trade chocolate. Mars Inc. will certify Malteasers in the United Kingdom and Ireland in 2012, and has already stated that by 2020 all of its chocolate will come from sustainable sources. The exposure of child slave labor used to harvest bulk cacao in Côte d'Ivoire, the world's biggest producer, will likely speed certification in the biggest cacao producing regions in the future.

As the demand for fair trade has increased, so have the number of certifying organizations creating a complex and confusing array of international options for businesses to navigate. For example, the Fairtrade Labelling Organizations (FLO), established in 1997, are the worldwide fair trade standard-setting and certification organizations. Fair Trade Certified, from Fair Trade USA, started certifying coffee in 1999 and has since become the leading third-party certifier of fair trade products in the United States, guaranteeing consumers that strict economic, social, and environmental criteria were met in the production and trade of cocoa. UTZ Certified, a Dutch nonprofit organization, established its Good Inside Cocoa Programme in 2008, after the success of its program for ethical coffee trading. Rainforest Alliance may focus certification on sustainability in its Sustainable Cocoa Program but wages and treatment of workers are also paramount to its mission statement. The Max Havelaar Foundation launched Max Havelaar cocoa in 1993 to guarantee small farmers in developing countries a fair price for produce. Today, even some companies have their own certification programs, such as ADM's Socially and Environmentally Responsible Agricultural Practices program.

On the face of it there seems no reason to quibble with this explosion of certification. Fair trade strives to "do good." Its ideals are admirable; third-party certifications are designed to ensure transparency and equity in international trade through sustainable development and better trading conditions and rights for producers and workers in underdeveloped growing regions. According to the International Cocoa Organization (ICCO), a fair trade-certified cocoa producer must comply with the certifier's list of social, economic, and environmental requirements and follow the organization's defined labor standards and conditions. At Fair Trade USA, for example, this means allowing the organization to audit and certify "transactions between US companies and their international suppliers to guarantee that the farmers and workers producing Fair Trade Certified goods are paid fair prices and wages, work in safe conditions, protect the environment, and receive community development funds to empower and uplift their communities."

Given all this, fair trade certification seems like the way to guarantee the safety and well-being of all people who grow cacao. But the straightforwardness of the label belies the complexity behind it, especially for fine flavor manufacturers and growers.

Simply put, fair trade is not an absolute guarantee of the ethics, equity, and sustainability of a chocolate bar. Unlike nutritional analysis, "fairness" is not scientifically verifiable. And just like genetics, it has no bearing on how that chocolate tastes. Kit Kat bars may now be fair trade certified but they taste the same as they did before. Moreover, fair trade does not require disclosure of how much chocolate is actually in a product, the origin of that chocolate, and how much of the rest of the ingredients are fair trade-certified. Is that Kit Kat bar a fair standard of comparison for a 67% fine flavor, single-origin chocolate bar with no inclusions, and beans bought directly from the farmer? What exactly is being compared?

This lack of clarity is one of the reasons keeping fine flavor chocolate producers and manufacturers from fully embracing fair trade certification. An even bigger reason is cost: fair trade certification is expensive—up to $10,000 a year. This is not so costly for big manufacturers. But for small manufacturers, it represents a substantial hit to their bottom lines; for many farmers, it is more than they make in a year.

In all fairness, certifying organizations have moved to address this. The Institute for Marketecology (IMO) program Fair for Life is a "brand-neutral, third-party certification programme for social accountability and fair trade in agricultural, manufacturing and trading operations" that seeks to complement existing certification systems by adding a layer of social and environmental responsibility and an understanding of smallholder needs. At least one fine flavor company, Madécasse in the United States, has started to work with them. In 2012, Fair Trade USA presented a draft of its *Independent Smallholders Standard* defining "the requirements for participation of groups of Smallholders (small farms that are predominantly family-run) in fair trade certification through their partnership with a 'Market Access Partner,' usually an exporter, processor, or NGO that partners with the group of Smallholders to commercialize their product and helps them to organize as a path to empowerment." But these solutions still leave behind complex problems as partnerships often dissolve when NGOs leave (and with them the funding for fair trade).

Of course, it is reasonable that the costs of certification by fair trade groups are high, given the remote locations in which cacao is grown. Moreover, small farmers who work three hectares of land or less produce 90 to 95 percent of that cacao. Even when those farmers are linked formally or informally as groups or cooperatives, only the cooperatives are centrally located. The farms are not, making certification of those farms nearly impossible and even certification of the cooperatives a cumbersome and time-consuming process.

Yet only half of Fair Trade USA's revenue is used to cover the personnel expenses to certify on location, and that includes collecting and managing fees, maintaining operations, and paying salaries and expenses for staff needed to process the astronomical amount of paperwork required in the certification process. This leads to the question many chocolate manufacturers and producers ask: Who profits most from fair trade? For example, Fair Trade USA may be a nonprofit organization, but it had nearly $10 million in revenue in 2009 (according to the latest financial statement available as this book went to press). How much of this really gets back to the farmer? Not enough, according to many in the fine flavor industry. Cacao is and will remain a labor-intensive crop to manage, harvest, and handle postharvest—more labor intensive than any other fair trade-certified product. Yet on top of the annual fee, a fair trade organization takes in more money per ton of cacao than the farmer does: about twenty cents per pound or $440 a ton for its stamp of approval. A farmer is lucky to get a quarter of that premium.

Even when the farmers are organized into cooperatives, as more and more are every year, then money for the beans goes to the certified cooperative and not directly to the farmers. Fair trade charges cooperatives the same as it does any organization, up to $10,000 per year for certification, plus $750 a day for travel expenses, but there is no clear standard for what each cooperative gets charged nor any standard for ensuring that the farmers are being treated fairly by that cooperative—or even if the farmers wanted to join that cooperative in the first place. In addition, that fair trade premium goes to the cooperative rather than the farmer, and the cooperative is supposed to use it for a community project (e.g., a new truck). Moreover, that cooperative of small farm holders isn't necessarily the exporter of the beans, adding a layer between the importer or manufacturer—somebody who goes to customs, puts the cocoa on the ship, and takes responsibility for it. If the cocoa has already been processed into liquor at origin or nearby before shipping, there are even more layers between the farmer and the certifying agency. Who profits from all this is more murky than transparent.

Then there is the problem that cacao-growing economies around the world are not uniform. According to the *CIA World Factbook*, Mexico's estimated GDP is $1.185 trillion, while Madagascar's is $9.4 billion (and its GDP per capita is still more than fifteen times larger per capita). Yet both are considered developing nations. But for fair trade, it is one-size-fits-all even though the wages and costs are very different. Not that what farmers make goes very far in building their futures in either country. According to Rizek Cacao, the average fine flavor cacao farmer in the Dominican Republic, who gets above-average yields due to the absence of disease there, nets $2,500 a year from a three-hectare farm that produces 1,500 kilograms of fine flavor cacao. If fair trade cannot ensure that the farmer earns much more through the 10 to 15 percent premium for fair trade certification—if he gets only, say, $60 of the hundreds of extra dollars those beans command, and everything else goes to the brokers and the middleman—what does fair trade certification mean for the small farmers who grow most of the world's cacao?

In the opinions of many throughout the industry: *It simply means we certify poverty.*

To most people in the fine chocolate industry—many of whom want to support fair trade certification—this is appalling and unacceptable. They believe genuine fair trade for the future means putting more in the pocket of the producers. Thus, most fine flavor chocolate manufacturers, especially those who hire, work with, or buy directly from farmers, exceed the fair trade standards in the country of origin and are calling on the entire industry to do so as well. They already pay more than the fair trade standard—and that's in addition to the differential paid for flavor beans. To certify at the fair trade level would be to *decrease* the price paid.

That's why Jorge Redmond and Chocolates El Rey did not go for certification. "We decided to go the route and say that we use 'fairly traded' cacao beans direct from small- and large-scale growers in Venezuela and always pay above-market prices for our cacao beans," Jorge says. "No middlemen or 'coyotes' take big cuts and seek to pay the lowest possible price to growers."

In the end, fair trade has its boosters, detractors, and people at every point in between. In the fine flavor world, believers see it is an important step to bridging the ethical divide between what's happening on the farms and what we believe is happening. Gary Guittard is big on certification and believes that ·in ten years all cocoa beans will be certified, not just because of pressure from customers but because it will help raise awareness of all the issues, labor included. Gary knows fair trade can't do it all, but it can help ensure transparency and serve as a gut check as businesses grow. The needs of small businesses may indeed result in new certifying groups that change the way certification is done. After all, small businesspeople know that as you grow, you lose touch with the jobs "on the ground." If a chocolate business manufactures in Europe or the United States, its eyes and ears can't be halfway around the world in Java or Madagascar, so it hires companies, managers, and technicians at origin. The biggest, most admired companies in the world aren't immune to the problems that often arise: Look what happened with Apple and the manufacturing of its iPad in China.

In the world of fine flavor chocolate, this separation is often inevitable. "If you are a large manufacturer or chocolatier and need one hundred tons a day, you can't be really picky about the beans you get," says Steve De Vries. "You look around and try to find the best that you can get, but at some point you just have to buy what you can get one hundred tons of and make chocolate. On the flip side, if your growers are getting one ton per year on their farms and you need one hundred tons, you don't have the time to deal with a hundred or two hundred farmers to get your beans. You probably can't even deal with the people the farmers have to deal with, but the person that deals with the people who deals with the farmers. That's three degrees of separation from the growers."

Given there is no single agency to expose the truth of what is happening on the ground, no industry voice for education to prevent people from saying whatever they want in their marketing, fair trade could fill the advocacy, information, and transparency void. Instead, the harshest critics see fair trade organizations as unethical opportunists who figured out how to take a cut of every pound of coffee or cocoa when there are already brokers and other middlemen involved. They see fair trade as a "hamster wheel" and "nonsense" (several call it a "scam" that gets you coming and going) that originated to improve the working conditions on huge coffee plantations and now puts thousands of dollars in the pockets of certifying authorities or government bureaucracy on the backs of cacao farmers and businesses big and small. Many say even when they can afford it, they would rather give those thousands of dollars directly to the farmers than pay fair trade to come in and tell them what they are already doing. Finally, fair trade critics dislike that there are no accommodations for the different needs of different communities or countries, and that the overall lack of transparency in fair trade means that consumers, by and large, have no idea what they are buying into—too many believe everything not fair trade is not fair.

The contradictions in this last point are not lost on Art Pollard: "Certification is a big business and makes an immense amount of money compared to the people buying and growing the products. And yet a lot of the customers who are insisting on fair trade are anticorporatists."

It is then poignantly ironic that fair trade has become a great name and trademark. And so most fine flavor producers and manufacturers, even if they do not embrace fair trade, acknowledge its power as a marketing tool. Those huge increases in the number of fair trade chocolates and coffees on the market are happening because fair trade certification is important to more and more customers and retail stores (such as Whole Foods in the United States), forcing more and more growers and bean and liquor buyers who specialize in flavor to comply in some way. (Belgians prefer Rainforest Alliance, according to Hugo Hermelink, who owns FINMAC farm in Costa Rica, which was also the first in Latin America to get certified by UTZ. The majority of chocolate produced by Italy's ICAM is now certified fair trade or organic.)

Consumers everywhere like what the certifications claim to say about a company's ethics and the transparency of its supply chain. They are willing to pay for that peace of mind and may get cantankerous when they can't—just ask David Castellan of Soma in Toronto about the nasty e-mails he gets about not selling all fair trade. Consumers seem to pay for a little false hope, too: One recent study conducted by the University of Michigan, reported in the *New York Times* and elsewhere, showed that people who ate "ethically" produced chocolate not only felt virtuous but that the halo effect extended to their waistlines: they believed fair trade chocolate had fewer calories.

A label helps you feel *less* guilt about what is for many a guilty pleasure? Now, that's the brilliant marketing! Which was exactly why Jorge Redmond was on the phone with the fair trade people in the first place. He hardly thinks the ideals of fair trade are stupid but he echoes the concerns of many in the fine flavor chocolate industry that in practice, certifications often offer more hype than hope. "In the end, it is about marketing. It is all hype. When [the fair trade group] discovered that we pay twice the international price directly to the farmers, they wondered why we needed a fair trade ticket in the first place. 'You are already doing that,' they said. And I said of course we are, but people want the damned ticket—the stamp on the package."

The marketing side of this story is for the next part of this book, as are other certifications like organic. The question at this point in our journey is: what to do to help those farmers? As Massimiliano (Max) Wax, vice president of the cocoa business unit at Rizek in the Dominican Republic, says, "Obviously the industry cannot close their eyes and behave like a Don Quixote because Don Quixote died poor, beaten, and battered in the end."

At the same time, the industry cannot be Sancho Panza— Don Quixote's sidekick—who simply tells the story. Fine flavor businesses know they cannot just make promises and assume transparency from what someone else tells them. They must educate themselves and consumers that there is not one size (i.e., label, business model, flavor) that fits all for cacao or cacao farmers. And to do so, they must get involved and literally get their hands dirty to, as John Kehoe, a former buyer and now vice president of sourcing and development at TCHO Chocolate in

San Francisco, says, "go beyond fair trade." They need to change the paradigm and work as directly as possible with growers at the source.

That's how, to paraphrase Margaret Mead, this small group of thoughtful, committed flavor citizens can start to change the world.

Pay Attention!

Margaret Mead believed that studying and working with people far away would help Americans and all people in the industrialized world understand themselves better and create a more humane and socially responsible society. This ideal resonates deeply with many in the fine flavor chocolate industry. The question is: does paying attention to farming and farmers from the ground up make for good business?

Paying attention to workers does have a long history in achieving gains in productivity in business. For example, in the 1920s, the Hawthorne Works in Chicago, Illinois, commissioned a study to test the productivity of its workers on the factory floor. Would the employees work differently in more or less light? The answer? They became more productive in more light . . . *and* less light. Productivity had nothing to do with the amount of light, but that someone was paying attention to light and thus paying attention to the workers themselves.

There seems every reason to believe that attention paid to cacao farmers should have the same desired effect and help preserve, protect, and propagate fine flavor cacao in the future. Turns out, however, just paying attention is not enough, nor was it enough at the Hawthorne Works: Productivity slumped once the lights stopped changing, once it was clear that no one was paying attention. Decades later, the Hawthorne study is held up as an example of social engineering, not social responsibility. Long-term gains required more than simple manipulation of the conditions that might have had more to do with fear than care for the workers; it required long-term relationships to sustain the gains.

Something similar happened when people stopped caring about fine flavor cacao in the early twentieth century and it was replaced by low-flavor, higher-yield varieties, and then again in the late twentieth century when people stopped buying a lot of cacao after production exceeded demand, and depressed prices and disease left a lot of cacao farms abandoned. Today, demand and prices for all types of beans are only going up. But supply is expected to fall off, especially for flavor beans, not only because of the threats of disease and other environmental factors such as climate change (nicely summarized in "The Future of Chocolate" in the February 2012 *Scientific American*), but also because farmers are again abandoning cacao. Supply and demand may be in line right now, but with future demand expected to increase by more than one million tons by 2020, there will likely be a general shortage of supply of fine flavor beans as higher-yield clones replace them. More significantly, there could be a general shortage of all cacao as farmers choose to grow other crops that pay them better and faster.

In other words, you can't preserve, protect, or propagate something from the ground up that does not exist on the ground. And chocolate manufacturers cannot pay attention to cacao farmers who aren't there.

In Europe, Felchlin, ICAM, and Barry Callebaut are not worried about supply short term but are concerned long term about supply as China comes online. In this regard, Barry Callebaut is particularly concerned about farmers dropping out in Africa. "In the Côte d'Ivoire, our biggest supplier, we are seeing more and more farmers stop growing cacao," says Mark Adriaenssens, Barry Callebaut's head of R&D in the Americas. "They go to rubber mainly. The world will need more and more tires now that China is almost the number one car market. You might not expect that in cacao. But it is the reality. The entire industry will have to work actively in improving everything, as directly as possible, and we think Barry Callebaut is one of the key pillars in the strategy in creating sustainable cacao because it is our lifeline. But we will not succeed alone in getting the million new tons needed in the future."

Will Southeast Asia help? Reuters reported in 2011 that Indonesia's chocoholics are leading an Asian quality wave. But at the same time, smallholder farmers, who own about 90 percent of the total plantation area in the third-largest cacao growing country, are also switching to rubber as well as coffee and palm oil, which can offer bigger profits. "The next three to five years, no farmer will want to plant cocoa and some will replace [it]," said Zulhefi Sikumbang, chairman of the cocoa producers' association Askindo. (Except in this quote and unless otherwise noted, cacao in this book refers to the *Theobroma cacao* tree, the fruit it produces, or the fruit's seeds; cocoa or cocoa beans refers to those seeds once fermented and dried.) Sikumbang also runs his own 1,200-hectare cocoa plantation in West Java. According to Reuters, in an effort to combat the decline, the Indonesian government launched a $350 million program in 2009 to boost cocoa production to more than 600,000 metric tons within five years but "it has yet to show results, with bad weather, failure to follow advice on planting techniques, and mismanagement working against the campaign."

Papua New Guinea could grow much more cacao but farmers keep rotating through crops from vanilla to coffee—whatever the speculators tell them will get the quickest and highest price. Vietnam might not be much help for the future, either. Dr. Pham Hong Duc Phuoc, director of the biotechnology center at Nong Lam University, loves the environmental benefits of cacao but knows growing it is directly connected to economics for the farmers: "The branches and leaves of cocoa grow fast and require frequent pruning, which is why the ground of a cocoa farm will be covered with a thick layer of leaves and branches. This layer of plant residue puts cocoa way above other plants in erosion prevention, increasing water-holding capacity of the soil and increasing soil fertility, which is very beneficial environmentally. Cacao also gave us a chance to introduce a new plant into the farming system of Vietnam and create new opportunities for income and reduce risks for farmers of Ben Tre. But none of this matters to future sustainability if productivity and pricing lag behind other plants." Which is why Vietnam is also switching to crops like rubber, coffee, and palm oil for bigger profits and less work.

Jump the ocean to Bolivia and you'll find, as Volker Lehmann reports, coca along with a little CCN-51 have become far more attractive to farmers than fine flavor cacao production. In Ecuador, fine flavor cacao also competes for its future with CCN-51 and other high-yield clones that are good for production but hard on the environment with their constant demand for water and fertilizer. But again, those aren't the real competitors. If so, Africa would be the "enemy," not soy and corn, which are highly productive and actively pushed by large companies. Every year, more and more farmers chop down their cacao trees and welcome global companies that give them seeds and chemicals and then, of course, pay them more than they can earn for cacao. "What a loss for the environment and the industry. We need to wake up," says Santiago Peralta of Pacari in Ecuador.

Compared to these and other crops, cacao is not only environmentally sustainable but is a model fruit for intercropping on new or existing farms (cashew trees serve as a natural umbrella for the cacao plants in Vietnam just as bananas do in the Americas) and can even help stem the loss of rain forests. In the case of rain forest deforestation, the environmental benefits are inarguable. But cattle are easier—and also in high demand, and so they are filling clear-cut farms and forests-turned-pasture in Brazil and Mexico and Costa Rica and . . . Yet even cattle don't compete in flavor regions like Trinidad and Venezuela, where production has gone down and down as oil has gone up and up.

Move north into Grenada and, according to Mott Green of Grenada Chocolate, cacao peaked in the 1950s because Grenada's people changed the way they live: "One of the reasons they have so little cocoa compared to what they used to have is that so many of the trees have been cut down to build houses. Grenada has extremely well-educated and intelligent people who are keeping the place going with other jobs. Grenada also has tourism. So cocoa has gone down and down and down and is just starting to come back because of the demand for world cocoa, and because nutmeg is dead from hurricanes and that was the main crop."

That is not the case in Central America, where cacao still grows in remote villages and farmers now realize they can grow bananas for a lot less work and have them to port in a few days. Passionfruit and pineapple are popular, too, and both are higher yielding and less labor intensive, and command better prices than cocoa. And this is before factoring in the devastating impact of frosty pod rot.

All this is exactly why Robert Peck of the World Cocoa Foundation says, "It's difficult to tell a farmer, 'Mr. José, you need to plant cocoa and you need to provide for your farm and your family the livelihoods that you can somehow make with these three hundred kilograms per hectare a year of your cocoa.' That's not something that Mr. José is going to do for very long. And for sure his kids are not going to do it. And at three hundred kilograms per hectare, no one can make a living. I do not care if it's fine flavor or if it's certified. No farmer can make a living on that. The discussion should be how are we to help farmers become more productive. How are we going to make them more successful in what they do? We don't want cocoa farmers to abandon cocoa. We want farmers to think of the trees as a valuable possession."

Robert's point about ensuring that the children of cacao farmers think the trees are a valuable possession ties directly into one of the most oft-quoted concerns about farmers and the future of fine flavor cacao: The average age of cacao farmers is around fifty-six years old. Truth is, that metric is a little deceiving. Ed Seguine of Mars Inc. explains it nicely: "Farmers are not aging. They have always been the old guys. And they will always be the old guys and the reason that will be the case is it's the old guys who own the land and they haven't died yet. Do you not understand inheritance? That said, it's not just that farmers are old; it's that these farmers do not want their children involved in cocoa farming. And the children do not want to be involved in cocoa farming."

Simply put, the fine flavor industry needs to engage the current *and* the next generation of farmers—to get them excited about the farms and their potential as viable, sustainable, and profitable businesses. A small step in that direction goes back to where we started: bringing farmers and their families chocolate made from their cacao. That's making it real, connecting to what it means to millions and what it could mean for them and their families. If they connect to the crop, they might consider its value and possibilities next to rubber or even higher-yielding clones. They might not abandon farming completely and move to the cities and find a terrible job for wages no better—and sometimes even worse—than cacao farming. But planting trees that take three to five years to see a return on investment, versus corn or soy that a large company with little interest in cacao will pay you to grow *now*? Never mind, as Steve De Vries notes, that corn or soy could be inter-planted with cacao, providing shade for the young trees and an annual harvest for those first three to five years. No wonder farmers don't want to grow cacao anymore— even older farmers who love the trees and the land. And for sure their children don't have the time or inclination to make that connection to the land and listen to talk of a brighter future in a decade. All they see is what their families are going through: hard work, year-round, for little money.

"The young generation hates farming. We're all getting older in this business. Even the new manufacturers like me are all in our forties. That's the biggest problem of all and it is a very scary thing," says Mott Green. "So what would happen ultimately if Grenadians don't start growing more and refurbishing more trees is that eventually big companies would come in and take over. Grenadians are very resistant to selling their land but everybody has their price, and a big company could come in and just buy the estates and own all the farms. And then they've got the labor problem here that they had in Africa and everywhere else in the world when this happens. No problem at all! They would solve it the way other people solved it: bring in the cheap labor—the Filipinos and the Guyanese. So now you have people who work really hard, from countries that didn't get lazy in agriculture, and your country has nothing. The people don't have their land. They don't have their jobs. And they are going to New York to try to make money because all the estates they used to have are being farmed by big companies with imported labor. I think that is a big danger in a lot of places these days."

And yet this is the classic story of "progress": Younger generations want to explore new things. Their inclination for the future is to move as far away from what their parents and the older generation did. Sometimes that means leaving the business entirely and giving up farming. Sometimes that means changing the products, abandoning cacao for another crop or clone. But perhaps that means changing the way things are done, learning about fine flavor cocoa, improving the production and process, and building a more successful business.

Clearly, it is that last possibility that offers the most promise to the fine flavor industry—both for the growers abandoning cacao and those who are growing it but living in poverty—and it is the hardest promise to deliver on. To solve an issue this big requires more than just turning on a Hawthorne-like "light" and bringing farmers chocolate to demonstrate that someone is paying attention. It requires a complete shift in the business model. Until very recently, global cacao supply chains were entirely based on an archaic business model of middlemen. The chocolate industry did not go out into the field. It depended on brokers. Manufacturers would call traders in Rotterdam, order 3,000 tons of cocoa beans, ask the price, and wait for delivery. Traceability and transparency were a joke. The brokers and exporters had the power to sell to the highest bidder and essentially say whatever they wanted about the beans they offered; stories abound in the industry of rotten beans, poor quality beans passing as flavor, and shipments and promised exclusives disappearing when another bidder swooped in.

Farmers are almost completely devalued in this model, and fair trade does nothing to pull open the proverbial kimono in cacao. You mean farmers and growers cannot survive on four or six cents on the dollar for their crops? Who is getting the other ninety-six cents? Hmmm, did that broker just buy a new BMW? And let's not forget there is usually another layer in this complex web that disenfranchises farmers: breeders (sometimes government-sponsored or owned) who grow and propagate seedlings to sell to plantations, cooperatives, or farmers. Too often, these seedlings are not only of poor genetic quality but also of poor quality, period. The majority of farmers just get the planting material that is given to them or recommended to them by seed gardens and, if the seed garden is government owned, that often makes things worse.

Welcome to the
Department of Licensing Bel-Red

E905

Please retain this ticket until your business is completed today.

Most services are also available online @ www.dol.wa.gov

6/23/2015 8:43 AM
Case ID: 6993247

So how does the industry preserve, protect, and propagate fine flavor cacao in a fragmented market where for so long the middlemen have traditionally controlled the product, the breeders have little sense of how the seedlings will perform, manufacturers have focused on production and profit over quality, and farmers are barely making a living?

That's where direct trade comes in. Direct trade can turn up the lights for the long term and ensure manufacturers are always paying attention and being accountable to the workers from farm to bar and bonbon. Direct trade has the potential to change the dominant model and create socially responsible and profitable relationships. "New manufacturers coming in are forcing the older ones to be more active and get into a relationship with their producers," says Volker Lehmann, a German national and farm owner in Bolivia who has worked in cacao for twenty-five years. "They are investing in traceability and also transparency. This is happening. This is a good thing."

What's the Fairest (Trade) of Them All?

How can the fine flavor industry best reach out to farmers? How can it find ways to enable them and empower them to experience a transformation that comes from a commitment to their communities? To answer these questions, the industry has an unlikely guide: the twelfth-century physician, philosopher, and scholar Moses ben Maimon, known to the Greeks as Maimonides and to his followers as Rambam. A source of wisdom for Jews, Christians, and Muslims alike, Rambam grappled with similar questions on matters of giving, and his answer was an exquisitely elegant eight-step Ladder of Charity (or Rambam's Ladder) that shows how we can live better by giving better.

Now, the business of fine chocolate is not about charity. But Rambam's Ladder is, at its core, about empowerment: The highest rung on the ladder is about giving freely to people who are in need, treating them with respect, and helping them achieve self-sufficiency. At the bottom rung of the ladder is the reluctant giver who throws money into someone's hand but avoids eye contact. This might be the manufacturer (or even a broker or buyer) who begrudgingly pays the minimum premium for the fine flavor beans as if tossing a coin in a beggar's cup without looking. The step is not to be disregarded but encouraged as a path to climb further. Think of it as the beginning of a relationship built on a more familiar cultural reference: "Show me the money," the famous line Tom Cruise screams in the movie *Jerry Maguire.*

The less one makes farmers ask for that money—the more freely the money is given—the better, but the point is also to pay more. Farmers need money now; cacao must provide a livelihood and show a profit before it can be a business that a younger generation of farmers wants to take over and transform. For sure, direct trade can be stressful and costly for those who live thousands of miles away from the farms, but it is important for quality control *and* it turns out the bottom line: Direct trade makes it easier for manufacturers to pay farmers more by bypassing the tangled web of traders, brokers, and exporters. In fact, without middlemen taking a cut, manufacturers can actually pay farmers more yet ultimately pay less for the beans. Everybody wins, except for the middlemen. That's the reason even giant companies like Wal-Mart, pilloried for so many of its bargaining and labor practices, have explored the benefits of the direct trade model; Wal-Mart started by buying some of its apples in the United States directly from regional farmers yet keeping the price paid for those apples the same as or lower than when it used middlemen and expanded to engage small- and medium-sized farms internationally.

There are complications, of course, beyond the stress and added hours required for direct trade, especially when you are not Wal-Mart and can control, if not own, the entire supply chain—especially in cacao and chocolate when that supply chain is so layered and complex even without middlemen. For example, fine flavor manufacturers may be willing to pay well above top-dollar prices directly to farmers for their highest quality beans, but even in the best-managed operations not every bean turns out great. What happens to the other beans? Who buys them? Do they have no value? Can direct traders help find buyers for those beans? Important questions. The hope is that for now at least, paying more directly to the farmer forms a foundation of trust and transparency to work together to find solutions.

Volker Lehmann experienced this firsthand when he "discovered" the legendary Bolivian wild cacao that Felchlin turned into Cru Sauvage. "I said I needed much more money for this project and Felchlin began to talk about the market. They were concerned the price was too high. But in the end, we agreed I would open up my books and show them exactly what I spent and this is how much I have to earn and this is what you need to pay me. They agreed in the end. It was a real open and transparent deal." Felchlin for its part even paid this generosity forward. According to Felix Inderbitzin, Felchlin's cacao sourcer,

"One year, instead of sending Christmas presents to our customers, we just gave the money to the farmers to buy books for the school. So maybe a farmer can afford to send the kids to school because they don't have to pay for the books. That's not about a bean, that's about social responsibility. We are helping make a farmer's life a little bit better."

These moments of generosity help the industry understand how deep the needs of cacao farmers go. For example, money for books is a wonderful gesture, but what if the children won't need the money for books because there is no school, or what if there is a school but there are no teachers? Caring about people in the supply chain must build on these singular moments of charity. Economic transformation must go beyond simply paying more than the premium directly to the farmers and extend to bigger questions of social responsibility on the ground.

It must also help farmers do their job better, as Robert Peck of the World Cocoa Foundation argued. After all, manufacturers can pay more and prices will rise when demand outstrips supply, but if there is not enough supply to meet the demand, the farmers will still not generate enough income to improve their livelihoods. As previously noted, cacao farms are small and have often been poorly maintained for years. Yields are horrible. If a farmer earns only a few hundred or even a thousand dollars more through direct trade, that's still not going to do much for the farmer. No price in the world right now is going to make it a viable, let alone interesting, business to his children. According to Steve De Vries, cacao traded at $6,000 a metric ton in the high price days of the 1960s and 1970s, which translates to an average of approximately $40,000 a metric ton today. Okay, that might do it. But in reality, to be a sustainable business at today's prices, productivity (while avoiding the dangers of overproduction learned in the 1960s and 1970s) is key. We must help farmers increase their yields from the standard 300 to 400 kilograms per hectare by teaching good agricultural practices or as Robert says, "help cacao farmers be cacao farmers again."

This is already the goal of a program targeting the conditions in the bulk cacao fields of Africa, to which one of the biggest worldwide givers, the Bill and Melinda Gates Foundation, and more than a dozen chocolate companies have pledged more than $90 million to fund: the WCF's Cocoa Livelihoods Program. Its stated goal is to improve the livelihoods of 75,000 Côte d'Ivoire smallholder farmers by 2014 and eventually 200,000 cocoa-growing households across Côte d'Ivoire, Ghana, Nigeria, Cameroon, and Liberia by improving "the quality and quantity of their crops and [providing] them with reliable opportunities to sell their crops so they can build better lives for themselves and their families."

Yet even if the Gates Foundation and big chocolate manufacturers like Barry Callebaut and Mars succeed in helping farmers get bigger and better yields from existing trees in Africa, even if more superclones arrive to take up more real estate in South America, resulting in short-term yields that are more than ten times current production—new problems arise with infrastructure. Many farmers will still not have access to the roads and facilities to transport and process this increased production. "Unfortunately for cocoa lovers, cacao is not grown near the sea or near the roads," says Robert Peck. "Usually it is up in the mountains or down the road and in difficult places, and harvested by gatherers who bring it in for processing." Without this infrastructure, costly bottlenecks will inevitably result and potentially damage the harvests.

The mostly small fine flavor manufacturers do not have the kind of money required to solve these big infrastructure problems, but they can help farmers become more productive and better at processing in established and new fine flavor regions. Deep, direct long-term relationships built on direct trade between the chocolate makers and the growers are critical in this regard. They are also how the fine flavor industry will achieve the highest rung of Rambam's Ladder: Giving farmers the wherewithal to become self-reliant through shared wealth, work, and partnerships.

This Rambam-like ideal is not surprising when you consider that a number of manufacturers got the idea for their businesses while working for programs like the Peace Corps, and they retain a connection through their businesses to the core values of those groups and to the value of personal connections. Fair and direct trade will never shrink your waistline, but social responsibility is huge part of the value added in fine flavor chocolate. And make no mistake, fine flavor chocolate will be sustained as a value-added product—value that comes from more than flavor but also a dedication to social responsibility through economics, community building, environmental sustainability, and stories about the farmers, process, ingredients, and history that go into every bar.

Manufacturers are thus increasing their efforts to ensure that their customers know their companies' relationships to farmers are not just words on a chocolate wrapper. For example, in the United States, Askinosie Chocolate puts pictures and names of farmers it works with on the front of its wrappers, while

Taza issued its first-ever annual *Cacao Sourcing Transparency Report* in 2011. The Taza report documents how it delivers on the direct trade and fair compensation promises it has made since 2006 and will reinforce those direct relationships through visits to the farms at least once a year. Taza also promises to pay a premium of $500 per metric ton above the New York Board of Trade price. The first report includes information such as how long Taza has purchased the beans from its sources, the most recent visit to the farm, who made the visit, and bean count. Alex Whitmore, the cofounder of Taza, calls the report "all about transparency and fair compensation," and the company complemented its release with third-party certification through the USDA-accredited Quality Certification Services. Taza even flew in some farmers from La Red Guaconejo in the Dominican Republic for a party at its factory in Massachusetts to celebrate the report's launch.

Duffy Sheardown of Duffy's Red Star Chocolate in England could not agree more with these kinds of approaches: "If I want people to pay more, I want them to know I connect to the farmers directly." Thus Duffy is one of several manufacturers trying to take direct trade and farmer advocacy to a collective level. He joined Art Pollard (Amano); Gary Guittard (E Guittard); Santiago Peralta (Pacari); Mikkel Friis-Holm (Friis-Holm Chokolade); Franck Morin (Morin Chocolate); Martin Christy (Seventy%.com); and others for a meeting hosted by Frank Homann (Xoco) in Honduras in 2012 to form the Direct Cacao Group. The organization is, according to Duffy, designed to "speak with a singular voice within the industry and outside the industry to improve the lot for cacao farmers and provide a coherent voice for independent chocolate makers." Its mission is a commitment to pay an above-average price to farmers for cocoa beans, work with them to increase the quality, keep the supply chain as short as possible so that as much of the extra price goes to the farmers (hence the "direct" name), and improve the percentage of fine chocolate sold overall. To commemorate the meeting, they planted two cacao trees on the beach where Christopher Columbus landed on Guanaja in 1502.

Those cacao trees are perfect symbols of hope for the movement of smaller chocolate manufacturers to make a real change on the ground and take a holistic approach to direct trade that goes beyond price. The trees could be viewed as a reset button for the entire fine flavor industry.

"Pushing the reset button is a great way to say it," says Frederick Schilling, who founded Dagoba and is now a partner in Amma in Brazil and Big Tree Farms in Bali. "This is a rebirthing process that is happening right now in the industry. Small flavor-oriented companies are pushing and recreating the chocolate markets around the world and engaging with the larger companies to make the changes bigger. We go to the WCF conference and talk to guys from Mars and Hershey and Nestlé and really talk about sustainability and fine flavor and getting them to come on line. It is very, very small, unfortunately. The intention is there. I don't think it is the intention of these larger companies who create this supply chain to keep it archaic. It just is a complex system that takes a long time to change. It takes a lot of effort. Those of us who are able to and have the time to go to origin—that's where the change has to happen—spending time there, shifting the business model and how it works. A lot more has to happen. We are these little mice who are scurrying around the world doing these little things and that is changing the entire industry."

Some Model Behavior for Farmers

What follows is a tour around the world of some of the ways those "mice" are building a more flavorful future and making fine flavor cacao sustainable from an economic, environmental, and socially responsible perspective. (Note: This next part is the longest single journey we will take in this entire book. Not that this is ever a bad idea, but please make sure you have extra chocolate nearby.)

Frederick Schilling is certainly putting his money where his "mouse" is. After selling Dagoba Organic Chocolate to Hershey's in 2006 for $17 million, he took some time off and "stewed in the lab." A few years later, he was "ready to reemerge, take my lab coat off, put my clown hat on, and have fun again." Maybe that's why when we first told Frederick that we were writing a book on the future of fine flavor chocolate and asked what he thought, he laughed and said, "Wooooooo."

Pressed to explain, Frederick asserted that it's an exciting time to be in fine flavor chocolate—no preconceived notions of what absolutely works, no agenda for all beyond do well by doing good for all people involved and make it a tastier world. But Frederick does have an agenda for his own companies: building on what he did at origin with Dagoba through Amma, a fine flavor chocolate made for Brazilians in Brazil, and Big Tree Farms in Bali, which in addition to producing nonchocolate products, currently specializes in a raw chocolate.

Big Tree Farms is the first bean-to-bar manufacturer in Southeast Asia and is built on direct trade with the region's smallholder farmers who grow a surprisingly high-quality cacao. While many Indonesian farmers have replanted clones and lost much of the high-grade Criollo that migrated there after the Spanish brought it to the Philippines, genetics on Bali were pretty much left alone, as the farmers weren't interested in replanting (as a largely Hindu and Buddhist culture, farms are not designed for sustenance, not income). Big Tree Farms is working with farmers in Bali and in Sumatra, while Frederick has also planted an ancient Java Criollo plantation in Indonesia to increase supply in a few years. In fact, all sourcing of ingredients—including the ancient local sweetener, coconut sugar, which Big Tree uses instead of cane sugar—is done directly and transparently and all production is done at origin. Big Tree Farms prides itself on this, marketing to customers how the company works with thousands of farmers throughout Southeast Asia and has "social equity built into the products' DNA."

Of course, for Frederick, as for so many others, that is "where the industry is going," but to get to where it wanted be as the first Southeast Asia bean-to-bar manufacturer, Big Tree farms needed to think beyond the bar. When the company could not keep up with demand for its bulk chocolate, it built a factory, and now that factory has attracted as much attention as the chocolate being produced there—maybe even more. Normally, a company would look at a steel frame and concrete structure for a factory, but Big Tree Farms decided to dismiss any preconceived notions and build it out of bamboo, one of the world's most sustainable building materials. Completed in late 2011, the factory covers 23,500 square feet and stands three stories high, topped with a thick grass roof. It is the largest commercial bamboo building in the world and a model for both economic (it was half the cost of a steel factory) and environmental sustainability. Not surprisingly, it immediately caught the attention of the *New York Times.*

"Our factory embodies what we stand for, sustainability, thinking outside the box, not accepting the status quo as how you have to do things," says Frederick. "Let's push the boundaries of what we can do and prove to the world that it can be done. Let's just try it and see what happens. So we are building a full chocolate factory in a bamboo building and Southeast Asia's first bean-to-bar factory. I know it sounds crazy but we are. It took some crazy thinking to do it, too."

In Brazil, the growing model could not be more different than Southeast Asia's—massive family plantations—and neither could Frederick's company, which is a farm-to- bar model that seeks to make Brazil's best artisan chocolate. At Amma, this can be done because the experience runs deep. His partner, Diego Badaro, is a fourth-generation cacao farmer, and he owns three cacao farms with Amazonian genetics that date back to the 1700s. But the operating principle is not simply to make great chocolate, but to use fine flavor cacao as a reforestation and forest preservation tool. The rain forest that stretches from northern Brazil down to Rio is 11 percent of its original size. But where it exists, the *cabruca* growing method thrives. Cabruca simply means that the cacao is planted within the forest—lower production but environmentally correct. It requires some clearing but the upper canopy of trees is still very heavy.

For centuries, cabruca helped maintain the forest and preserve it . . . until the witches' broom arrived and the cacao disappeared along with the forest, cleared for cattle and corn. "It is just a shame, a global shame, this deforestation," says Frederick. "That's why with Amma it is about promoting chocolate as a method of forest preservation. Brazilians are a very passionate people, very nationalistic. They have a lot of pride. When they learn about what is happening with their forests and how cacao could play a very key part in preserving the forest and replanting the forest, they respond. That's what our message is and what we represent in Brazil. That is very tangible for Brazilians. Is that going to be tangible for someone in Norway or New York City? Not as much. So, at Amma, the farm-to-bar model is built for the people in Brazil. But both the Big Tree and Amma models play a very important part in making a great chocolate. It all depends on what business owners want to create and how they can add value to what they make."

At Pacari in Ecuador, Santiago Peralta appreciates Amma's focus on national pride but wants nothing to do with the big plantations or productive clones that dot his homeland—what he calls "an industrialized way of thinking," like deforestation of flavor. "All the really fantastic flavors are in Ecuador. The whole world appears in front of you through this chocolate. It is not just because it is my country. There must be a chance for these beautiful flavors to exist. I want the world to understand this. I want to share it."

Pacari is now ten years old and a model for what a small locally owned chocolate company can do, launching landmark work on organic, raw, and biodynamic production. The company has direct connections to 3,000 families in Ecuador, something that would be incredibly time consuming for a company the size of Pacari if it were not based at origin. "It *is* easier for me because

I am in Ecuador and from there," says Santiago. "But it still all comes through connection to the farmers and talking to them enough to understand their needs. Most of the people buy from co-ops or a farm that buys from a group of farmers. Even though there's a lot of goodwill here, some people say they are doing it and don't. How many of us are really connected with farms? We're just a little bit deeper into it. We can be transparent and guarantee quality in Ecuador that others cannot."

For Santiago, the most important thing is to save what still exists, and that comes first from paying more to the farmers he works with, and extending that model of higher prices for better quality cacao across Ecuador, undoing problems like mixing beans. Pacari serves as one example of a solution: "Quality through proper harvesting, proper fermenting, proper drying, and proper processing, and then make the people understand that this is special chocolate. If I'm doing that I am sure of two things: I am making an investment for the entire industry because people are interested in trying new chocolates, and I am increasing the price paid for the cacao in Ecuador by five or maybe ten percent because we are getting a lot of fame and now people want to copy us by making a nice flavor like a Pacari. That's great. That can potentially put millions of dollars a year in the pocket of farmers. That is what we need to do not only with Ecuador but with all the farmers in the cacao regions."

One of those farmers, Hugo Hermelink, owns and runs FINMAC, a plantation of 110 hectares and almost 140,000 trees in Costa Rica. His work inverts the Pacari model by managing directly from farm and farmer (not owner to farmer) to his manufacturers at origin. Unlike Santiago, Hugo was not born and raised at origin (he is Dutch), but he has deep roots in Costa

Rica and has maintained his farm and flavor in a region that used to be famous for them. His father originally raised cattle before he started planting cacao on the farm in 1985. Hugo moved from Bolivia to take it over in 1994 and has focused on improvements from better genetics for flavor and disease resistance to high productivity to organic certification (not easy when commercial banana and pineapple plantations close in your fields on either side) to build his business. He may battle frosty pod rot, higher labor costs than in other countries, and insects, but his beans thrive in the hot temperatures and rainy jungle conditions.

It is the first two of those problems—disease and wages— that Hugo calls a "killing combination" for Costa Rica because the cacao requires extra management to maintain and harvest, increasing his already high labor costs. Thus he is trying to take some of the labor out of the equation. "What we have tried to do is find a way of mechanizing and of getting higher yields per hectare, and getting better qualities and better price—a whole package of actions to survive," says Hugo "I think we've found ways to mechanize at least part of the harvesting and part of the processing of the harvest. We need to do this because as a company and as a farm, we want to get more done in the same workday so we pay high wages to take the pain out of that workday. Efficiency is everything. To offset this we need something special, so the first thing we focus on is better yield and mechanizing of the harvest, which means there is less work per ton of cacao."

Most of Hugo's cacao is turned into cocoa liquor for his biggest customers in Belgium and a few smaller ones like Steve De Vries and Gail Ambrosius in the United States, but he has recently expanded into that farmer-to-bar model so he can get better prices for the processed product. Hugo adds value to the product by having local artisans gather his trees' fallen leaves to make paper and boxes (Gail uses them for her products) and to the environment by welcoming diverse animal and plant life (his plantation is the site of a sloth study). He also pays his success forward by sending some of his liquor and giving some of his land to the Amazilia cooperative: eight local women who make Amazilia Chocolate Organico bars and bonbons, which they market and sell themselves to contribute to their families' incomes. The project has been so successful that in 2012, the women will get a grant from the Costa Rican government to set up their own chocolate equipment and expand internationally. "I think you'll see much more from them in the coming years," says Hugo.

The Amazilia cooperative, of course, is not the traditional grower cooperative found in the cacao trade. Of these traditional cooperatives, the largest in Costa Rica is the Talamanca Small Producers Association, with more than 1,000 producers who grow more than a third of the country's cacao. This scalability of size and the ability to link several cooperatives into one network—many with an increasing eye on manufacturing, not just growing—is one of the advantages of the increasingly popular and progressive cooperative model in the world of fine flavor cacao.

At the smaller end of the cooperative spectrum, you have producer-owned cooperatives like Mott Green's Grenada Chocolate Company and El Ceibo in Bolivia. Grenada is more egalitarian and the larger and more experienced of the two when it comes to fine flavor, processing about twenty-three tons of cacao and producing about 250,000 bars a year—small compared to overall production on Grenada, but unique as it produces the only cacao not regulated and exported by the Grenada Cocoa Association. Like Hugo with FINMAC and El Ceibo, Mott had the foresight to get organically certified earlier than most—at the same time he formed a cooperative of cocoa farmers. There are now ten farms in the Grenada Chocolate cooperative and everybody is a shareholder. All this, coupled with the Grenada Chocolate cooperative's willingness to put a high value on the beans, has made it more and more popular with farmers. And with the cocoa market expanding and half of the cocoa trees in Grenada abandoned under bush, it offers the possibility of sustainable living for even more farmers as their farms get rehabilitated to reach 100 percent productivity (a process that can take just three years).

According to Mott, who hails from the United States, Grenada Chocolate is still the only cooperative model in cacao that is a pure cooperative from farm to factory with no foreign investors at all. But is that sustainable and replicable? "I think that it is possible and sustainable to have a small cooperative to do what we are doing in the right circumstances with the right people, and with chocolate that is good enough and marketed well enough to command a high price. But this is never going to be a model for everyone because we are so pure in our cooperative setup—because I am an idealistic activist. Because I make everyone, including myself, take the same salary. They're not fantastic salaries, but they are way better than average, especially for some of the work. The fact that the senior chocolate maker and I also get the same salary is not as fantastic, as we have much more responsibility and some people call me a socialist because of this. There is a kind of Marxist way in believing salaries should be equal because a day's work is a day's work. But I am not a socialist. I am not a capitalist either, obviously. If anything, I'm an anarchist."

The biggest challenge faced by this egalitarian cooperative model is its greatest strength: fairness. With no increased salaries to offer, the cooperative cannot afford to pay someone to take responsibility for what Mott has been doing for ten years and let him focus on growth and marketing. But Grenada offers huge possibilities for what Mott believes is among the most delicious cocoas in the world, and because of its setup it is also a very ethical chocolate to eat. He also realizes that no matter what he does, the world gets its Grenada cocoa anyway. "It's for sale. It's available. The big boys like Valrhona and Chocovic have big contracts and have had them for very long time. What we are trying to do is create small boys right in Grenada and to grow that as fairly as possible without external investors, but that makes profits very marginal because we are so idealistic."

El Ceibo in Bolivia is not strictly egalitarian in its cooperative model—it simply promises fair and equal distribution of income and dividends to its matrix of fifty autonomous cooperatives—but it is trying to go one step further and take a person like Mott out of the production equation. El Ceibo is "tree to bar" with everything owned and run by the cooperative and no foreign shareholders. Its board, management, and employees are from Bolivia. Founded in 1977, it began producing semi-finished products in 1980, converted to organic farming in 1987, and soon exported its first certified organic cocoa to Switzerland. Three years later, they started producing products for the Bolivian Market and had more than twenty items in the market—none of them fine flavor. It wasn't until 2008 that El Ceibo decided to produce fine flavor bars to capitalize on the international demand for fine flavor chocolate. The problem was that without a "Mott" they had no idea what they were doing. They had great Bolivian genetic material but no idea how to make fine chocolate or how it was supposed to taste. To solve this problem, they asked an international expert to consult: Chloé Doutre-Roussel (known in the industry as "Chloé Chocolat").

"For two weeks, I did an analysis of everything: cacao, fermentation, drying, the machines, and mainly the people, because these people wanted to enter the international market, which is extremely competitive, and they were just cacao growers," says Chloé. "I wanted them to taste the chocolate and understand the market so I brought them around twenty fine chocolate brands including Cluizel, Valrhona, Marcolini, and Domori, as well as some other rubbish, so they would understand. I told them about marketing and strategies and where the trends are going and we worked together to come up with the choices so they could make a decision. Everything was transparent. Everything was honest. Everything we did, I wanted them to be able to do without me. It would take a long time but if I died tomorrow, they could continue. It remains today the only chocolate that is made one hundred percent by cocoa growers."

Of course, full ownership from farm or bean to bar is not necessary for a cooperative to thrive. Ghana's now-famous 45,000-member Kuapa Kokoo cooperative (*kuapa kokoo* means "good cacao farmers" in Ghanaian) does not make the chocolate at origin. Its stake in manufacturing is through its partnership in Divine Chocolate in the United Kingdom and the United States.

The cooperative does command a higher price for some of its best beans, but its farmers are always guaranteed a premium on top of the price paid and can decide at the village level how to use that premium.

In Madagascar, the American company Madécasse Chocolate uses partnerships at origin and has created a kind of hybrid: a chocolate company that works directly with cooperatives and farmers, a manufacturer, and other companies. Madagascar is unique because, by virtue of its legendary flavor, it is completely a supply-and-demand market unaffected by stock markets. Farmers grow it; someone will buy it. Madécasse understood this and created a model that was featured in the *Wall Street Journal* on how small businesses are setting up innovative models in Africa. "You could call it a middle partnership—vertically integrated but not completely owned," says Brett Beach, a partner in Madécasse.

The arrangement has huge plusses and minuses. "The big plus is that everybody does what they're good at and does not try to do everything. The farmers and cooperatives we work with grow cocoa well, so you let them do their thing. You have a production partner that does what he does well, and he's always there and his family has been there for generations. And we are Americans who lived in Madagascar [as part of the Peace Corps] so we can kind of relate to this and hold everything together while we think about what new flavors will work and what will appeal to the market. The admitted weakness is that we don't own the cocoa farms and we don't own the factory, but all the equipment is there. The cocoa farmers thus have the right not to work with us. We could invest for years teaching them, and somebody could come in from another country and say, 'You know what? I'm going to pay you more for a month' and take it all away. But we think the strengths are greater."

And they have been tested. Brett and his partners have survived a coup, a severe allergic reaction to chocolate developed by their taste tester, and embezzlement by an employee. Still, the benefits outweigh the pitfalls. "We can say all the benefits are built into the product before it ships. When you buy our exotic pepper bar, we know exactly how much money stays in Madagascar because we know what we pay for all these things," Brett adds. "You can argue all you want about the value of paying people to print and package and that they're not cocoa farmers but employees, but there's no arguing dollars and cents of what stays in Madagascar. And those dollars and cents should be the scientific hard proof of how a company operates."

Jorge Redmond of Chocolates El Rey in Venezuela faces a different political reality: the government of President Hugo Chavez makes life difficult for any private companies there no matter how they treat their workers. In fact, El Rey has sought to balance the inequities found in conventional third-world trade by establishing Aprocao, a democratically run cooperative managed by El Rey. Its goal, according to Jorge, is to allow farmers to "create a good business that can be handed down to their sons and daughters who keep it going as a profitable business. Venezuelan chocolate is revered even though its output is small— less than 1 percent of the world's production. Premiums can easily exceed $1,000 above market price, but we are among the least productive countries in the world in terms of yield. With such soft yields, it doesn't provide a good income for an average family of five. We need to make sure that these farming families can have a good life by growing it the proper way."

To do this, El Rey has developed a shovel-ready, fifteen-year, $500-million plan for increasing production by five times the current amount through increased yields on existing plantations, plus 25,000 new hectares for growing cacao. But nothing is easy when dealing with the Chavez government. "Unfortunately, until now we haven't found any interface between the government and the private sector to do such a thing. The Europeans have a program that we are trying to get access to but with our government we are cut off from that. We've got access to the Bank for Inter-American Development but we can't get the money without the okay from the government. We've been waiting over thirteen years to get this thing going. All the challenges here are fundamentally political."

And things are potentially looking worse and will only continue to until Chavez is replaced with a differently minded government. According to several sources, including the BBC, Chavez has the cacao industry in his sights. His government has started seizing land, creating bureaucratic tangles to control the movement of all cacao in the country, and is working on a cocoa processing plan (with money from Cuba) that includes a factory and school and could lead to nationalization of cacao like coffee and oil, making it impossible for growers to make money—at least in the long run. In the short run, as Chloé Doutre-Roussel reports from a recent visit to Chuao, farmers told her that they were paid seventy bolívares by the government instead of the forty paid by an Italian manufacturer. Propaganda? Perhaps. But Chloé, Jorge, and others believe that this is all part of the government's plan to cut off the private sector's supply and reduce exports of cacao—eventually, in Chloé's opinion, exporting only liquor, butter powder, and chocolate with "more added value for the people of the country."

El Rey has already seen one of its most promising projects vanish. The company created a farm as part of a larger partnership with a petroleum company in Barinas, where Chavez was born (that may have been its biggest mistake), and developed an Ocumare-type hybrid that had wonderful flavor and aroma. "Basically it was an unbelievable product," says Jorge. "We developed a section of this farm to grow it and we got yields of 800 to 1,000 kilograms per hectare—four or five times more than average. We were beginning to develop a mechanism with our growers to have a population per hectare that would give yields of around 3,000 to 5,000 kilograms per hectare. And then the farm was taken over by the government and destroyed."

El Rey did manage to preserve some of the breeding materials and is working on a much smaller project with the beans that should see yields by 2016 if the government does not get in the way. "In general, the cocoa industry in Venezuela is geared to helping growers," continues Jorge. "They are our best customers and we are their best clients. So we are doing a lot of things together and we are defending each other from the government. We are of tired of being on the crappy end of the stick. The government has us up against the wall. We have fifty-six steps to export now. It is a very hostile environment for the private sector. But we try not to give up."

Francisco Gomez, chief operating officer for Casa Luker Cacao's global business unit in Colombia, must be thankful his company faces none of the governmental problems of its Venezuelan neighbors, otherwise its efforts to build sustainable fine flavor cacao supply chains and long-term relationships with farmers surely would have stalled. Casa Luker has been making cocoa and chocolate since 1906 and Francisco has been there since 1962. Since he started, Casa Luker has become one of the biggest cooperative-based companies in the world, selling its chocolate mostly in Colombia and to a few fine flavor customers like Anne Weyns of Artisan du Chocolat in England. The company is also one of the most adept at increasing yields from fine flavor trees by helping farmers break free of monocropping. "All the industry is talking about sustainability because of the need to preserve the postcolonial model that was implanted in Africa with the smallholder model," Francisco told the audience at the July 2011 Fine Chocolate Industry Association (FCIA) meeting. "But in my opinion and the opinion of many others, this small model of two hectares will never be able to produce enough income to feed a family. There is no way around it. I don't see what type of certification or funding or income from foreigners or any government program that could make a cacao farm this size produce a decent living."

Instead, the Casa Luker model deploys many of the same tactics discussed already—a mixture of research, breeding, technology, and training to increase farm production and size— but combines it with diversification to add value to the farms. As part of his detailed presentation to the FCIA, Francisco's charts illustrated that in the thirty years from 1965 to 1995, Casa Luker selected cacao trees with disease resistance and productivity, adding nonproductive wood for shade and plantains for additional crops. By 1995, farmers earned the equivalent today of more than $45,000 annually for the cacao, and also had wood valued at more than $9,000 and fruit valued at more than $11,000, making for a total of more than $65,000 annually. Over the next twelve years, Luker went even further, turning farms into agroforestry systems with plantains and productive wood, and grafting new elite cacao trees selected for good quality and high production. Farmers planted alternating rows of wood, fine flavor cacao, and plantains with the cacao trees that have sexual affinity planted near one another. Yields started lower, of course—only one hundred kilograms per hectare in the first few years, but then rose to 1,500 kilograms by year five, and 2,000 by year seven with productivity of up to 98 percent. The value with the wood and plantains: more than $83,000 annually. The current stage of the project focuses on even more efficient and productive flavor trees with the hope that gross farmer income could top $100,000 annually.

While these numbers seem high to some, no one disputes that the Casa Luker tactics have been successful at increasing farmer income. Casa Luker is also diversifying its planting models, balancing the difficult small-farmer work with the expansion of its plantation into what may become the biggest, most diverse fine flavor plantation in the world: 500 hectares in northwest Colombia with an agroforestry system incorporating fine flavor cacao, plantain, and wood species. Does its plantation expansion spell doom for small farmers and the cooperative model? Likely not, given the projected increased demand for cacao, that mechanization can only go so far in increasing efficiencies and reducing labor costs, and that there are between six and seven million smallholder farmers growing cacao in the world.

In fact, Rizek Cacao in the Dominican Republic is expanding its smallholder model and buying land to increase the size of the small farms to five hectares—the size of a cacao farm the company believes it needs to be for a family to survive whether a member of the cooperative or not. Of course, the farmers may be smallholders, but Rizek is a big, high-end organization with a tremendous number of employees and farmer partners, and much land. They are one of the better examples of large corporate investment in a commodity. When they do something like experiment with solar drying techniques, the scale and investment is tremendous. Is that a bad thing or a reason to be suspicious of their ethics, as some are? Of course not.

Max Wax and others at Rizek are smart businessmen and see cacao as a business and a livelihood for farmers—exactly what many in the fine flavor industry say it needs to be to appeal to the next generation. Rizek may not have an egalitarian cooperative structure like Mott Green's in Grenada, and it may have historic family ties to banking and shipping, but none of that makes it or any company like it inherently bad. Cocoa, like any industry, is bound to have good and bad versions of its business models. Cooperatives are no exception, so questions should be asked of *any* cooperative or company proclaiming direct ties and selling stories about its work on the ground in cacao on its websites, brochures, and packaging: What are the problems faced by the cooperative and farmers be they governmental, environmental, economic, and bureaucratic? If there are problems, are they being addressed? Are the promises of farmers sharing in the cooperative benefits being delivered? Are local farmers being forced to join the cooperative in any way? Answers may be difficult to come by but no answers will ever come if these questions are not asked continually.

After all, the majority of the manufacturers and sourcers chasing flavor around the globe cannot afford to set up shop at origin. Many bean-to-bar manufacturers, like David Castellan at Soma in Toronto, are deeply dedicated to quality, diversity, and sustainability. They will pay more but still mostly buy the six to eight bags of beans they need from different regions through brokers and produce remarkable chocolate from those beans. But as more and more embrace the bean-to-bar model and as those bean-to-bar companies grow, the hope for most in the industry is that they will embrace direct trade, paying a premium directly to the farmers, spending considerable time at origin, and expending extra effort to make a real, lasting connection. They'll need to. Constant maintenance of relationships with growers is essential, because no relationship—even a direct one—can ever be strong enough when so many farmers live in poverty. A farmer under pressure to sell to someone else who has more money—be it another small company, a large chocolate maker, or a broker—is often going to do it. As Duffy Sheardown of Duffy's Red Star in England says, "Finding anyone to sell small quantities of beans is the main problem. I will always pay more for top quality cocoa beans, and as I try to buy direct, I can pass some of the normal middleman profit back to the farmers. But I need to have people who will sell me the beans."

Visiting at origin as much as possible helps guarantee delivery, quality, and transparency, and bridges gaps in knowledge and culture, especially in less developed or established flavor regions. That's why newer, small bean-to-bar manufacturers around the world, from Michelle and Dean Morgan of Zokoko in Australia to Shawn Askinosie of Askinosie Chocolate in the middle of the United States, despite having established strong reputations very quickly in their growing markets, know they must devote considerable time at origin.

"Tiny manufacturers and family businesses like us, we have limitations," says Michelle. "I do think there will be increased pressure in the future only because of the number of people, many much larger than us, who are trying to source it directly. We are playing in a bigger game and the pressure because of the increased demand will raise more issues of exclusivity; certain chocolate makers will want to tie up a particular region or growing location. We are at the mercy of what happens at origin on the ground. That is part of the game that you are playing when you're working with an agricultural crop like coffee or cacao. That's why Dean is working so closely with farmers on the ground. He blames me for all his gray hair at the moment."

For example, Zokoko works directly with a cooperative in Papua New Guinea, adding value by doing everything from helping with the physical work like drying the beans to taking the chocolate back to the cooperative, which Michelle called "a powerful revelation" for the farmers and a foundation of their relationship. This also means they are almost back at where we started: closing the cacao circle of life by tasting the chocolate. But in Papua New Guinea, like other regions, this circle often has unavoidable layers beyond the farmers—something a manufacturer like Zokoko can only navigate by spending time at origin. "It comes down to understanding the internal politics," says Dean. "I learned that from the coffee side. The main exporter controls the shipping line so you have to get them on board or you won't get product out. They have to see that there is money in it for them as well. And if farmers don't have the money to get it out, what was the point of picking it? In Papua New Guinea, sixty percent of the crop rots on the tree. We need to make sure that we take responsibility to get the product out. So we have shipping and logistics partners, too. Now our farmers get a better price based on quality at their location and they don't have to worry about logistics. That is why it is essential to take the time to understand the local situation, even if it is just for the handful of growers that we can work with."

Shawn Askinosie would agree and believes the nearly two years it takes to see results at newer origins like Tanzania and the Philippines are an absolutely inevitable and critical byproduct of creating direct-trade relationships: "The idea for me, from the very beginning, was to find farmers who I could trade with directly. That presents a set of challenges that people who buy from brokers don't really have, especially as I didn't start my search with a preconceived notion of having a bean from Madagascar or from fill-in-the-blank because it is a great flavor quality. I thought, 'Where can I find farmers that I can trade with?' I didn't lead with my bean; I led with, 'Where can I find a group of farmers?'"

Like other manufacturers, Shawn is flying under the radar in these less-familiar origins, where big companies and brokers are not buying large quantities, but he is still paying the thirty to forty farmers he works with in each region a price that is far above the fair trade standard. He is investing time in helping them grow, manage, and process higher-quality beans for their whole area. He doesn't just put the faces of farmers he works with on his packaging; he profit-shares with them.

This profit sharing, however, comes with a cautionary tale for other manufacturers who think this kind of money model changes everything: "People think that profit sharing with farmers has really changed their lives," says Shawn. "No, it hasn't. It is just the right thing to do after five or six years together, but it hasn't really made a difference in terms of what we get from them. Sharing in the outcome of the success motivates employees in America. But it is not quite that way in the rest of the world. What do you mean, they don't want to make more money? That's the American way. But that is not necessarily the only way farmers in Tanzania or the Philippines think. They want better for themselves and their families but it is a mistake for Americans to think that they think the way we do."

It would also be a mistake to think that because they are living in poverty, these farmers are unhappy and inhospitable to foreigners. "Tanzania is just another world but I've never experienced anything like it," Shawn says. "One of the things that is patently obvious and uniform across all of the places and farmers I work with, and at the same time has been shocking to me, is the level of happiness and hospitality that I've encountered. It has been uniform, and I would say that the people in Tanzania take it to the highest level that I've ever encountered anywhere. And yet they have the least."

So what's the value-add for a company like Askinosie in these underdeveloped areas? Goodwill. For example, in the Philippines, he understood the irony of buying food from a community where he saw the hunger firsthand, so he started a sustainable food program with an elementary school that had several kids on a malnourished "watch list." Shawn helped the parents make a product called *Tableya* (a traditional hot chocolate drink), which he bought for $1 per unit and sold to American customers for $10. All the profits fund the food program. Well over 100,000 meals have been provided, all of them through sustainable commerce, not donations. Planning is underway now for a similar program in Tanzania. "Hopefully this will strengthen the relationship that we have with the growers and increase their loyalty to us," says Shawn. "Hopefully they will care about us and develop a long-term relationship."

The Askinosie program is exactly the kind of innovative—albeit small—community model that will help preserve and protect flavor in the future. Shawn calls it *kujengana*, the Swahili word "for building each other up." Kujengana certainly invokes the spirit of Rambam's Ladder, as well as Margaret Mead's belief in the power of small groups of citizens to change the world. But whatever it is called, it is all made possible because of direct trade with the farmers. That's what most people in the fine flavor industry believe is the foundation for the future—not to imitate any one model but find a "flavor" that works and explore all the possibilities as directly as possible so the industry can grow.

In fact, for some companies this long-term relationship is *built* on deeper explorations of flavor, which leads into the last stop on this part of our expedition on the ground: fermentation and drying.

Fomenting Fermentation for the Future

When we asked John Kehoe at TCHO if he thought there was enough flavor cacao being carefully grown and fermented to meet the demand of artisanal chocolate makers, his answer was swift, to the point, and reflective of most in the fine flavor industry:

"No, I think there's a lot of flavor cacao being badly grown and badly fermented."

Of course, this is not the case everywhere. An experienced farmer like Hugo Hermelink in Costa Rica knows that fine flavor means more than just genetic testing or genetic selection and adjusts his fermentation and drying process to produce a more even product. In countries like Venezuela and its villages like Chuao—where Art Pollard saw the generational continuity at work—cacao has been fermented for hundreds of years, and fermentation can be remarkably precise. But in flavor regions like Peru, there is very little history of fermentation and drying and most farmers know little about the process. In the past, they picked the cacao with the coffee and left it by the side of the road for a middleman to buy at a low price. If the middlemen did not come? They would throw it away. There was no feedback. There was no way to communicate what the end result had the potential to be.

John Kehoe knows this Peruvian story and others like it as well as anyone. He has spent two-plus decades in the industry, first as an exporter of cocoa beans and then sourcing beans from farmers and exporters to get manufacturers the highest quality he could find. He knows how much things have changed since he attended a meeting in 1998 at the American Cocoa Research Institute (before there was a World Cocoa Foundation) to talk about sustainable cocoa, and a lot of the veterans in the room had never been to a plantation. According to John, that was the beginning of people getting on the ground and looking at sustainable methods by integrating with environmental groups or talking to tree crop scientists. Ten years later when he joined TCHO, he had the opportunity to go even further in developing reliable sources of cacao for the future by "putting development back into research and development" and paying attention at origin, specifically during the postharvest process with its TCHOSource program.

Now, TCHO may treat chocolate more like rocket science (no worries, according its cofounder's bio, his former company made parts for NASA), but like many modern manufacturers, the company still understands that genetics are just the beginning. John Kehoe calls this "the human factor" and what "brings potential to life" from the attention paid in tending the cacao through harvesting, fermenting, drying, selecting, and bagging the beans for export. Simply put, working directly with a farmer, cooperative, or plantation to grow the finest beans in the world is great, but if that farmer, cooperative, or plantation doesn't ferment or dry those beans properly then the beans won't perform well or command a premium. It doesn't matter if a bean is pure Ancient Criollo or CCN-51; genetics won't matter if fermentation and drying destroy or damage the flavor. Soon enough, consumers will learn that those off tastes in a high-priced chocolate bar are not its artisanal qualities coming through but the result of beans that have been over- or underfermented. (For those unfamiliar with the fermentation and drying process, see the short summary in the back of the book.)

The good news is with proper treatment, the beans could get better. But a lot of work needs to be done. "There's a tremendous lack of knowledge in the fermentation and drying process at origin in places like Peru, or most anywhere really," says John. "Very few, if any co-ops, maybe a very few sophisticated plantations or exporters have some idea of flavor."

That's where a program like TCHOSource comes in. It helps producers earn a better living and get great beans in the same way TCHO's neighbors in Napa do with grapes: helping farmers grow beans by breeding high-quality varieties and grafting programs, and involving them in the fermentation and drying process. TCHO does it with cacao at origin by installing Flavor Labs—three in Peru, one in Ecuador, one in Ghana—that allow farmers to make small batches of chocolate with their beans and understand the flavor.

"When I came to TCHO," John adds, "I said 'How are we going to do this? How can a chocolate company expect the farmer or the cooperative to deliver the high-quality flavor when they don't really understand that?' I saw we had this fantastic lab here and a team that can build anything. So we took these labs and we installed them. We are teaching the farmers how to taste. Some of it is mutual self-interest, sure. It can't be 'we are an organic fair trade chocolate company and we want to change the world and do good . . . and by the way our chocolate sucks.' The point is if we are making a high-quality product, we need high-quality, consistent cacao. We need to, as a company and as an industry, invest in the relationship and bring technology to engage deeply with farmers about flavor and help them understand that the true value of their product is in its flavor. Not how many kilos it is. How does it look? How is it fermented?"

If all this sounds expensive, it is not. TCHO's Flavor Labs cost less than $10,000 to build. But building the lab doesn't mean the flavor will come. As John notes, "It is one thing to install a lab and another thing to teach people how to use it, and to train them in sensory analysis so that they can learn how to appreciate levels of astringency, acidity, bitterness, or pick up on flavor notes and then use a rating system so we can talk about levels of flavor intensities. The first phase has been getting the labs in. Now our work is training them how to use it and sensory development so that we have that common language and understanding of flavor."

In the Americas, there are several reasons for this lack of a common language and understanding. Some of it is infrastructure, both roads and physical structures that do not exist. Some of it is lost time, since many farmers stopped growing or caring about cacao when demand and prices dropped in the 1980s. Some of it is that big chocolate and candy companies can burn off any off tastes in the manufacturing process. And some of it is the lack of flavor experience: Much of the cacao is being converted into the locally popular drinking chocolate for the national markets, which has very low demands for quality—even unfermented cacao can be used in drinking chocolate.

In the parts of South and Central America, growers might try to treat cocoa beans like coffee, but coffee is fermented for hours and cacao is fermented for days. The complex flavor development and transformation that occur in the fermentation and drying process for cocoa beans are incomparable. That's if farmers can even get the beans to ferment. With small farmers producing the majority of the world's beans, flavor or otherwise, volume is an issue. Too little cacao and the beans will not generate enough heat for good fermentation. This can happen anywhere in the world. The Hawaii Chocolate Meeting in September 2011 discussed farmers' ongoing issues with small fermentation heaps of one cubic foot and the need to ensure heaps of four to six cubic feet to get the cacao to ferment properly. Even in prized growing regions like Madagascar, farmers often just bring their ten or so kilograms of fresh cacao to their own little areas, where it will not generate enough heat to ferment well, if at all.

To solve the problem, Hawaiian growers discussed the possibility of farms buying and adding in fresh seedpods from other sources. But that mixing requires assurances that all the beans not only are the same quality but also ferment and dry on the same schedule—which is exactly the problem faced by fine flavor beans in Ecuador: everyone mixes. Santiago Peralta of Pacari and a few buyers in Ecuador may have tight controls on their supply chains and processing, but the rest of the country often mixes beans that do not taste or ferment the same way at all. With no premium to be had for Nacional, there is no incentive to keep beans separated—even if some are CCN-51. Some cooperatives have rules against mixing with CCN-51 and some refuse membership with farms growing CCN-51—a trend that will hopefully continue but can be hard to confirm countrywide. Nacional and CCN-51 pod shapes and colors may look very different, but once the beans are harvested, there is no absolute way of distinguishing the difference by sight.

In June 2011, Jeffrey Stern visited an Ecuadorian cocoa trader's "patio" and saw this mixing in action as small producers brought their beans to be bought by a broker on the spot or fermented and dried together and then held until a buyer was found. These buyers then consolidated the beans and sold them to exporters, a practice that leaves much to be desired in terms of transparency and quality control, especially to manufacturers who have no idea what the beans are or their exact origin. The

Ministry of Agriculture doesn't control the patios, so brokers can say whatever they want about chocolate and the beans being shipped. The growers who produce excellent cacao are powerless to get flavor premiums in this situation.

"There are some government and private sector initiatives to stop mixing," notes Jeffrey. "You go out onto some of the traders' patios and you'll see posters to this effect. The Ecuadorian Nacional Research Institute or INIAP has a small chocolate lab set up and they are trying to work with small farmers so that they can take the beans and make a small quantity of chocolate out of them. [Then] the farmers can learn about the factors that are contributing to the taste. Unfortunately, the institute has little to no idea how to make chocolate, so what they are making is not that great. Transmar, a commodity house, ran a pilot project to buy Ecuadorian cacao *en baba* (in the pulp) so they would know what they were getting from the very start and could ferment each type of bean separately. But their work is on such a small scale. Without the premium for Nacional, the movement will be lethargic. It will take decades to get the problem fixed."

Cooperatives can help solve the volume issue. They aggregate small harvests of beans and the farmers share the benefits of coming together. But too often, the people running the fermentary at the cooperatives have as little experience with or education on the process as the farmers. There are, of course, exceptions like Luker, Rizek, and Grenada, but they have decades upon decades of experience. In places like Ghana, buyers can't even get to the cooperatives that exist and instead buy the beans from intermediaries, dealers, or brokers. Felchlin only managed to improve the beans they bought at origin by working with

Ghanaian native Yayra Glover. His involvement has produced rapid results, largely because of fermentation and drying improvement. "Normally a project of this size needs three to five years, because you have to train the farmers and change their mentality. They're used to doing the fermenting and drying the beans the same way for one hundred years and now we are asking them to do this special kind of thing our way," says Felix Inderbitzin, Felchlin's cacao sourcer. "You have to change the farmers' understanding in order to get the good quality beans out of them. That is a long process."

Barry Callebaut, the largest chocolate manufacturer in the world, is moving in this same direction in the largest cacao-growing region in the world, Côte d'Ivoire. Leery of good-quality cocoa being poorly fermented and dried incorrectly on fires that leave a smoky taste, they will train farmers by opening five of their Cocoa Universities there. The schools will be centrally located in the prime growing regions and will teach small groups of farmers to be models for their communities both pre- and postharvest. "The hope is that these model farmers improving the yield and quality will inspire other farmers to follow and then the model farmers can teach the other farmers directly, not us," says Mark Adriaenssens, Barry Callebaut's head of R&D in the Americas. "It is a self-sustaining system where the farmers will see the results of the model farms and will try to copy them so they can have better income."

At least in the Americas, Ghana, and Côte d'Ivoire, there is some fermentation going on. In Queensland, Australia, the newest cacao-growing region, there is absolutely no fermentation history, so Michelle and Dean Morgan at Zokoko have been going north from their headquarters near Sydney to help. Meanwhile, north of Australia in Indonesia, the world's third largest grower of cacao, there is little current need for fermentation. Years ago, when its beans were being shipped off to Europe, fermentation was part of the cacao culture. But today the industry turns to Indonesia and the surrounding area as a main source of cacao for cocoa powder and butter. To make cocoa powder and butter you don't need to ferment even if it would help get rid of the pulp to speed drying. Fine flavor is not the point; rapid production at the lowest cost is.

As a result, when Big Tree Farms formed in Bali and looked to add value to the cacao culture there—value that could be passed on to the farmers directly in the future—"fermentation was step number one," says cofounder Frederick Schilling. "We teach the farmers how to ferment the cacao and then once we have the fermented cacao, especially organic cacao, we can sell it in a more specialized marketplace where we can receive some premiums. We have a very specific model where we do some profit sharing with the farmers on that step. We've increased farmers' income by thirty percent just through that basic step."

From his perspective halfway around the world from Big Tree, Carlos Eichenberger, the founder of Danta Chocolate in Guatemala, could not agree more. "The biggest problem we have here in Guatemala is postharvest," he says. "That's why when Art Pollard of Amano and Martin Christy of Seventy% came here, they left very disappointed. They did not find any well-processed cacao. The cacao itself is very nice. The fruit is large and productive but the fermentation, more than anything, leaves a bit to be desired. The farmers are stuck on using X amount of days instead of knowing it's ready because it's ready. That's been my biggest hurdle: going to the farms, training the people, and getting them to stop saying 'Okay, it's been fermenting for five days, let's stop' but to do the cut test to see if the cacao is actually fermented properly or not."

Today, the vast majority of the beans Danta uses are sourced in Guatemala directly from two main suppliers and two different plantations at which Carlos has trained the farmers on fermentation and drying. But there are still issues with other growers as well as the NGOs and the Guatemalan Exporters Association, to whom Carlos is trying to pass on technical knowledge and proper guidelines for processing cacao. Things are improving, but as in Ecuador, slowly. "I still have to send back several shipments of beans because they were just very badly processed," Carlos reports. "But there has been progress. There's been a lot of money spent on teaching and trying to get these people to actually process their cacao properly in order to find international buyers. That's what we have been drumming into them incessantly. And I won't compromise. I have established a bit of a reputation as being extremely picky and someone who takes extreme care of the product. It is their choice if they want to sell their cacao for fifty dollars a sack or two hundred a sack in the future. It all depends on how you process it."

And that is a long process to get right because true understanding of fermentation and drying in the future goes beyond infrastructure, mixing, and volume. It even goes beyond process. "For too long, most farmers who are fermenting and drying have been taught a 'process': this is what you do, not this is what you're creating," says Art Pollard of Amano Chocolate. "You follow these steps and if you turn the beans over each and every day for five days and sun-dry them for X number of hours each day someone will come and buy your cocoa."

Of course, just like wine grapes, fine flavor cacao is never the same from year to year. Weather and temperature changes not only affect cacao preharvest but can throw the age-old standard fermentation and drying formula off postharvest. The heat of the fire and speed of the drying process affects the beans, too. For example, drying beans more quickly will trap the acetic acid and create a more acidic chocolate that is a little fruitier in flavor. Understanding this requires a huge amount of education. And unfortunately, it has not been happening at the speed most people in the fine flavor industry need it to happen. All this is the ultimate reason why the fine flavor industry works so hard to establish itself at origin or visit or even own the farms, have the farmers sample chocolate made from their beans, and build open relationships based on mutual respect that allow the direct and free flow of information between grower and manufacturer without the farmer taking the education or reeducation personally.

"How can we adapt technology in the lab and bring it to the field, but also use it to bring creativity and empowerment to the very relationship with our farmers?" asks John Kehoe. "This is only possible to do directly. It is impossible to do as a broker. There are too many proprietary things and raw materials and strategic relationships that chocolate makers have—certain things they are doing with fermentation and drying that are very proprietary—and a broker's currency is confidentiality and reliability to deliver. It is incredibly complex and challenging, but hopefully this will bring new people in with new technologies and new ways of looking at it, and we are just at the beginning for everyone taking steps in this direction."

This is exactly why Frederick Schilling and others believe the advantage in the future goes to manufacturers based at origin in cacao-producing countries or at least those with the most established direct partnerships, especially in places like Brazil, where there are fourth-generation cacao farmers like Diego Badaro, Frederick's partner in Amma, and Joao Tavares, whom Frederick calls "a drying and fermenting alchemist." But even in less established regions, work at origin is proceeding quickly and opens up all kinds of experimental possibilities, such as Big Tree's work with cold processing to create its raw chocolate. (They will also expand to introduce a roasted chocolate by early 2013.)

"Fermentation is a critical step for flavor development and has barely been explored on a conscious commercial level," Frederick notes. "There is so much that can be done at this stage of the chocolate-making process. That's why manufacturers at origin have an inherent advantage over makers in North America or Europe. We can have multiple batches going on at the same time, each one with different fermentation variables: moisture, inoculations, temperature, and more. This brings chocolate making to a whole new level and hints at what we are doing for some of our future chocolates: fermentation experimentation that will be able to be made into microbatches very quickly and easily."

This is the direction many fine flavor growers and manufacturers are going: understanding how postharvest techniques can create two or three different chocolates with the same batch of beans from the same farm or farms. The variation from any source could be extremely broad, offering manufacturers and even farmers a range of possibilities. In fact, TCHO knows the producers they are working with are selling beans to two or three other manufacturers, each of whom wants the flavor slightly different for its customers. By understanding how that flavor is created, by tasting and experimenting on their own, the farmers could learn to adjust their processes and tweak flavor for those customers' needs. "They learn to do that and they will get their highest prices in longer-term relationships," says John Kehoe. "That will help create the fundamental ingredients in a sustainable business model: demand and price."

That is exactly why Volker Lehmann, who found and properly processed the legendary Bolivian wild cacao that Felchlin turned into that Cru Sauvage, is pursuing a path similar to Frederick's. He sums up the entire situation neatly when he says, "In the last 200 years, the chocolate industry was just looking at themselves. They were never looking at the farms or at the trees or the farmers. Nobody knew what was really happening in the process of it all in fermentation and drying. This is all a very recent development this trend of the past ten to fifteen years with people reaching out to find new sources. This has all given a push to more technique and better protocols for fermentation and also more insight into what is actually happening in the process. Today, everyone realizes that they should come down and talk to the farmer and explain it to him. And if the farmer doesn't understand, they should make a project to explain exactly what they want."

Volker knows exactly what he wants from his farm and farmers. He is developing a large commercial cacao plantation of about 1,000 hectares in Bolivia that will produce a mix of different qualities that can be customized for the chocolatier: "If the client wants a certain type of bean with a certain fermentation, we can do it. Somebody can say, 'I would like to have this type of cacao, can you do this?' And I will say, 'Yes. What are you willing to pay for [it]? You want Grand Cru cacaos where we ferment it perfectly and use the best blends of Criollo or whatever hybrids there are? Something less expensive?' We are looking at interacting with the chocolate industry where someone says, 'Hey, can you do it a little less fermented or a little more?' And I can do them both."

To do this, Volker has worked for three years with Anthon Berg Chocolatiers in Denmark to develop a technique that takes "the same cacao and through fermentation makes two different bars with two different tastes just by changing the fermentation technique." Volker actually invented three different techniques and is engineering them all to be more adaptable to matters of taste. For three years now, he has been making chocolate using these techniques, sending around samples, and then adjusting based on the feedback. And that feedback can vary wildly: "I sent samples last year to top US companies and I got mixed reactions but in the end sold the same cacao to Zokoko in Australia and it won three medals at international shows. I just sent the sample and said, 'Hey, here, you make good chocolate out of this,' and they did."

All these fine flavor people will also need to make sure consumers understand all the work on the ground and are willing to pay for it so the industry can grow across the board. "We have these fine flavor cacaos that people buy twenty bags of and it gets a huge story and a gold medal at some show. That's it," says Volker, hints of optimism and resignation in his voice. "We are talking very little volume and relatively low money. There's no big money in this. It's a prestige product."

"This should not be the case but it is."

"Cheap Chocolate" Should Be an Oxymoron

Reflecting back to the story that opened this section, and given the truth of what Volker said, it's amazing Art Pollard survived to bear the expense of experimenting with those Chuao beans. When Amano Chocolates was starting in 2007, Art attended a trade show to pitch his first products. At lunch, he found himself at a hot dog stand talking to one of the other vendors who was interested in his company's chocolate.

"I would *love* to use your chocolate to dip pretzels in," he said "How much does it cost?"

"Fifteen dollars a pound."

"*Fifteen dollars!*" he screamed. "I won't pay over three."

"Really? I can't even buy good-quality cocoa for three dollars a pound."

That pretty sums up the biggest disconnect between farmers, manufacturers, and consumers: Art Pollard spent more for his beans than this person was willing to pay for the finished product. This person did not see all the work being done from the ground up. Neither do most consumers. But they must. Chocolate may be a matter of choice for consumers, but it is a matter of survival for farmers—a delicate balance that does not necessarily work for long-term sustainability.

All chocolate lovers, but particularly fine flavor customers, must accept the fact that cocoa producers are underpaid as individuals and even in cooperatives—poor farmers working together in a group remain simply a group of poor farmers. Consumers *must* understand the amount of work that goes into producing quality cocoa if it is to survive in the future. That's what will create some sustainable livelihoods for these farmers. If it tastes fantastic, people should pay more. And if people pay more, farmers will find a way to produce more. We shouldn't need movies like *Nothing Like Chocolate* (2011) by Kum-Kum Bhavnani, but we do. Too often our outrage at media reports about forced child labor behind cocoa from Côte d'Ivoire and chocolate built on the backs of poorly compensated indigenous farmers and inexpensive imported labor outweighs our sticker shock when a Hershey bar goes up in price or the first time a potential fine flavor customer picks up a bar of Amano's Chuao chocolate. Ultimately, *both* the Amano bar and the Hershey bar are too cheap.

Unlike in the wine market, most chocolate consumers do not see the value proposition of paying more for quality at the high end and many have a hard time justifying fifteen dollars a pound for the world's best chocolate bars. In fine flavor chocolate and the industry as a whole, only chocolatiers and bonbon manufacturers have crossed this price threshold with consumers willing pay up to one hundred dollars a pound for bonbons. There are many possible reasons for this: packaging, in-store presentation, the fact that many boxed chocolates are given as gifts, the fact that there is far more refinement and artistry—not to mention quality ingredients—going into those bonbons. Perhaps their European heritage ties them to a lingering belief that the French, Belgians, or Swiss make the best chocolate. Yet even for fine flavor chocolatiers, the value proposition does not extend beyond the bonbon; they cannot charge much more than most fine flavor chocolate manufacturers for their chocolate bars. Forrest Gump never said life was like a *bar* of chocolate.

Simply put, most people think of chocolate as a commodity and not a food, and the reason goes beyond process and back to a lack of connectivity between consumer and farmer and the work that goes into producing a great bean before a manufacturer can even produce great chocolate. They don't see the people behind the flavor and the infrastructure, education, and maintenance of direct relationships that go into every fine flavor bean. They trust the packaging and their experience, at least in much of the world, of chocolate as an inexpensive childhood treat. Even chefs who willingly pay $6 a pound for perfect produce will balk at chocolate over $4 a pound. Then again, most chefs have at least seen vegetables grow. Few have been to a cacao farm. Truth is, few consumers consider the origin of the bean when eating chocolate of any kind. Even those who are wise to the thrill of a fine single origin bar or a bonbon from Patrick Roger or Michael Recchiuti, or who have stood rapt watching the magic of the machines at small manufacturers like Soma in Toronto, only see the final steps in the process. The fields are far away.

These are just some of the factors keeping cocoa beans the most highly undervalued fruit crop in the world. And it *is* fruit— just like coffee, which trades at a much higher rate and requires far less care and maintenance. Or compare it to nuts: cacao is one-third the price of pecans, less than half as expensive as hazelnuts, and cheaper than any nut except peanuts. And those crops can be machine harvested and require little work postharvest.

The conclusion is simple: as more and more people chase flavor around the world and turn it into the bars and bonbons we crave, it is absolutely crucial that cheap chocolate become a thing of the past. Things are changing, but they need to change faster. Until then, the big manufacturers—whose main concern is volume and keeping the price down—will govern the world of cacao.

"People need to pay more and we need to make the public appreciate this so they will pay more," says Santiago Peralta. "They need to know this is special stuff."

PART THREE

To Market, To Market: Craftsmanship, Customer Education, and Flavor

It is not the employer who pays the wages. Employers only handle the money. It is the customer who pays the wages.

—Henry Ford

A Nightmare

That's what Chloé Doutre-Roussel calls her first day as the chocolate buyer for London's Fortnum and Mason in 2003: a nightmare.

To many of us, the first day on a job at which your sole responsibility is to talk about, explore, and taste chocolate and then decide which of the finest ones to buy sounds like a dream come true. But it wasn't for Chloé, a respected chocolate expert known throughout the chocolate industry as "Chloé Chocolat." And not just because she had to wear a special uniform she disliked, because her basement office was windowless and had pipes running all over the place, or because she was being introduced to everyone as the "privileged chocolate star" from Paris, breeding resentment and hateful glances from the rest of the staff. The problem wasn't even that when she finally was taken to the chocolate counters, there were eight journalists and TV crews and a PR person barking at her to answer all their questions about a job she had started minutes before.

No, the day really went from bad to worse when the BBC invited an exhausted Chloé for a television segment on chocolate *that evening*.

"First, the PR people dress me like the Queen of England—things I would never wear in my life—and I'm completely depressed," Chloé recalls. "Then, I am pushed into a taxi and we arrive after the show has already started. That's when the producer tells me she has prepared a blind tasting for me. I tell her I never taste chocolate in the evening, but after such a long day even I am beyond caring. And then suddenly I hear them announce that I'm going on air. I am taken on the set and the host hands me a plate of chocolate and says, 'Now *eat!*'"

Chloé mimics the slightly withering look she gave the host that night as she continues her story: "I looked at her and said, 'You do not eat first. You *smell* first.'" Then, without ever taking a bite, Chloé sniffed each piece and separated the five chocolates into two groups, labeling one mass market and the other fine. She was, of course, correct. The stunned host wondered how she did could tell the difference. "I told you, I smelled it," she said.

That evening in 2003, a BBC TV host learned what Chloé and other international experts, manufacturers, and chocolatiers throughout the fine flavor industry repeatedly explain: The pleasure found in fine chocolate is not about eating or even tasting alone. True, there are important distinctions between munching mindlessly on a chocolate bar or bonbon and savoring its pleasures, but for Chloé, these pleasures go beyond taste. In fact, in a 2012 BBC radio interview, she explained the differences between mass-market and fine chocolate by connecting it to another sense—hearing—likening the best mass-market brands to the "boom boom" of a jazz drumbeat and fine flavor brands to the rich complexity of a symphony.

"You have to listen to chocolate in the same way that you listen to music," Chloé explains. "You have to listen to appreciate certain music. You have to give the flavor notes of chocolate many tastes and chances. It doesn't mean you'll actually like the chocolates but you may and you certainly will appreciate them."

Chloé Doutre-Roussel got into chocolate because no other product had ever made her happier. Every time she ate chocolate, she felt intense pleasure and wanted to understand how and why. A couple of times she was so completely shaken by what she ate—by Steve De Vries in 2002 and by Domori in 2003—she even lost the words to explain her feelings. It was not that those chocolates were the best Chloé ever had or perfect in texture or flavor (neither was) but they offered her a new experience—pleasure that she could not put into words because, she says, "it was totally unrelated to my past experience." She needed a whole new sensory vocabulary. It gave her hope for what was to come.

And, oh, what was to come: an explosion of small manufacturers and chocolatiers, especially in the United States and Canada but also across Europe and in emerging fine flavor markets like China, Ecuador, Brazil, Costa Rica, Bolivia, and beyond—all of them chasing flavor and customers locally and internationally. As a result, the choices in fine chocolate are almost as overwhelming as the possibilities. This has led many people to question whether such variety is too much of a good thing: is it economically unsustainable, given that demand is increasing and chocolate is so underpriced? Others are far more optimistic, seeing the dawn of a new model that has little relation to what has come before, a model that embraces diversity and consumer education at the local and global level.

Chloé's main concern is that consumers won't or can't appreciate this diversity soon enough for many fine flavor manufacturers and by extension, the diversity, to survive—though this is not necessarily an absolute negative. "What we need to do with chocolate is appreciate the differences between the two-dollar, five-dollar, and ten-dollar bars instead of just producing more and more chocolate the people will not understand."

This leads Chloé into her current nightmare: people who want to skip all this experience and education and know her favorite chocolates. "The first question you always get is 'What is the best chocolate—what are my favorites?' I'm tempted to say, 'Who cares?' Like humans, you have chocolates that you will like a lot, those that you like but that have some negative qualities but the positive overpowers the negative, and then some you just don't like at all even though they're very good. So it's a question of finding what positive things are most important to you. Like music or friends, each person must make his own opinion and those opinions evolve. Your favorites also depend on your culture, background, childhood, and your mood at the moment. I think that in our favorites we should always have the ones that we know please us and that we know we will get pleasure from, and those that shake us and we are not used to them yet but we eat them until our body is used to that new vocabulary."

Yeah, but who has time for that?

Manufacturers and chocolatiers will tell you they follow their hearts in creating their chocolate and bonbons but too many consumers follow the herd and buy into marketing and the power of packaging. Even those with a deeper understanding of flavor, cacao, and chocolate want what they want—dark or milk, organic or raw, meltingly smooth or crudo, nuts or nibs or perhaps pink peppercorns or some coarse Balinese salt—and want it now. Fine flavor manufacturers are selling to a generation that wants pleasure fast. "When most people eat a piece of chocolate we want that pleasure immediately: boom! That's the music of mass-market chocolate," says Chloé. "We do not have the patience to wait. Just like in love, we have lost that art of the flirt and letters and wooing and seduction."

So, can consumers learn to slow down, taste, explore, and value the costly complexity of fine flavor? That's the dream, of course, and there is hope. But consumers can be fickle and even dismissive when it comes to matters of taste, and though the future looks bright by some measurements, sometimes the numbers aren't what they seem.

An Educated Dream

The tasting room at Mast Brothers Chocolate is often filled with visitors to and people from their Williamsburg neighborhood in Brooklyn, New York. Rick Mast, who had worked with New York City legend Jacques Torres to learn how chocolate was made on a larger scale, likens the room to one you might find at a small winery—a chance for people not only to buy but also to explore and learn more about the craft itself, in this case craft chocolate. But unlike a winery, children are welcome to taste in the Mast tasting room and savor the possibilities beyond what they may be used to. After all, as Rick says, "theirs is the kind of optimistic feeling I hope will be propagated and associated with our chocolate."

So why do so many parents often stop their children from feeling this way?

"I find it very curious that when a mother brings in her daughter, she could be five, six, or ten, and the mom just lets her take some samples and start eating them, that ten times out of ten the girl will say, 'Oh, this is great, I want more. I love it, mommy!' But when the mom says, 'You might not like this; it's really dark chocolate . . .' Ten times out of ten that child doesn't like it."

To Rick Mast, these are the interesting problems to solve in the future, and the solution begins by stepping back and remembering not only how disconnected most people are from where cacao is grown but also understanding how and where chocolate is made. "The good news is that it's chocolate. It is never going to stink. People are already smiling when they are coming in and tasting it. And then the fact that they get to be connected to something more interesting and inspiring and fascinating, it goes from good to great. But those people that just walked past the window . . .?" Rick says, gesturing at the big glass windowpanes that open onto their Brooklyn neighborhood. "They don't even know there's a chocolate factory in here. We deliberately check ourselves in here to make sure we're not getting too caught up in our chocolate maker world. We are chocolate geeks. We are chocolate nerds. We have to get out of that world and think, 'What is it that we are really trying to do?' What I am trying to do is introduce craft chocolate to that ten-year-old girl whose whole life is in front of her and whose life I think would be better with a Mast Brothers bar than with a Hershey bar."

The majority of craft chocolate manufacturers and chocolatiers around the world think and feel the same way: Life is better not just with chocolate but also with flavor. Which is why "education" is the one word almost every flavor researcher, grower, manufacturer, and chocolatier uses when talking about how to get people to connect to, appreciate, and pay more for fine flavor chocolate in the future.

In fact, Mast Brothers—which benefits from considerable publicity opportunities by virtue of their singular position as a craft bean-to-bar manufacturer focusing exclusively on fine flavor in media-mad and food-obsessed New York City—does not have a publicity and marketing department per se; they have an education department, which is less a semantic difference than what Rick calls an "important mindset" for any chocolate company. That's because the Masts, like many fine flavor manufacturers, know that even their best traditional advertising, marketing, and publicity efforts can't compete with the big money of multinational chocolate manufacturers. For them, education starts at the exact opposite of "mass," with direct connections—tasting with and talking to as many customers as possible, which can sometimes lead to surprising results.

For example, Rick notes how much fun the company had with its Papua New Guinea single-source bar. "Essentially in the industrial chocolate world Papua New Guinea is sort of like a defect source. It has this smoked defect," he says. "And we thought what a terrific way to educate people of the different things that can happen on a farm and see how it impacts flavor. We thought that was a great opportunity and to our surprise it became people's favorite. That was one thing I thought for sure the people wouldn't like and we would really have to go into overdrive to convince people. But people ended up really liking it."

Steve De Vries learned this same lesson years ago as one of the small fine flavor pioneers in the United States. "If you are making basic chocolate, putting enough sugar in, then the majority of the world will go, 'That's yummy,'" says Steve. "But I go after flavors that aren't necessarily immediately yummy but are uniquely mine. It drove me crazy for years. First, I was doing it and people would taste my chocolate and just go, 'What the hell is this?' When I started doing more tastings and I got people one on one then I would do a lot better. So when we can't be there, we battle against immediately yummy."

Dan Pearson of Marañón Chocolate in Peru knows the pitfalls of battling the yummy all too well. "Many pastry chefs and chocolatiers have said to me something like, 'There's one thing about this chocolate you have, Dan: you can't put anything with it. You could put in whole cream and make a ganache but you can't infuse it.' Some of them say their business models are to buy the cheap crap and infuse it with artificial flavors—that's how they make their money. My chocolate is too expensive for their target markets. They can use the worst damned beans in the world infused with more vanilla and more soy lecithin and more artificial flavors so we're convinced that it tastes like something good. But the difference between that and the natural flavor of my fine flavor beans and other fine flavor manufacturers' beans? It's like the difference between fresh lemonade and Country Time."

These days, it is not only manufacturers in established markets like the United States, Canada, Japan, and Western Europe battling against the yummy and Country Time preferences of their customers and chasing customers for the future, but also those in newer markets, especially manufacturers in fine flavor-growing regions pursuing customers inside their borders.

Michelle and Dean Morgan's Zokoko in Australia may be, in Michelle's words, "tiny, tiny, tiny," but they are putting an outsized effort into tastings to grow their local market. "People are blown away by the fact that two chocolates can taste completely different even if they are at the same percentage," Michelle reports. "Then they want it finer and finer in terms of texture and realize that the Cadbury Dairy Milk that they had in their childhood is grainy and that the darker chocolates in the supermarket are still thirty to thirty five percent sugar and have milk solids. Then give them something that is dark but not necessarily bitter and not overroasted and they really like it." Michelle and Dean have even made a few batches from a couple of different growers in Queensland, Australia to good results and have great hopes for the future. "So it is all about education in Australia right now. On the ground, it is about helping farmers understand that what they do affects the end product—how what they do with harvest and fermentation and drying stages is so critical. With consumers, we are after the sort of light-bulb moment in which they try the chocolate and say, 'Whoa, I can't go back to stuff from the supermarket.'"

That is exactly the foundation Mott Green built his Grenada Chocolate on: "The local market is what's been keeping us going and sustaining losses in investment and export. That is been fantastic. Grenada is small so we can make the chocolate on the side of the island to bring it to the other side of the island and it's only twenty miles away and the cocoa is right here and all around us. There are tourists here to sell the chocolate to and locals here who have a taste for the chocolate and are not in abject poverty. That has worked and made us much more profitable."

Across the Caribbean Sea in Guatemala, Mott has been a tremendous influence on Carlos Eichenberger as he builds his Danta Chocolate. While Carlos has had some success in Europe and expects decent US distribution by 2012, he has survived despite exporting very little (only about 10 percent of his production in 2011) and selling into a national market dominated by the sweet drinking chocolate so popular in Latin America. Guatemala may be known as a coffee- and sugar-producing country but, according to Carlos, those crops were introduced later in the country's history; cacao is native. "That's the first thing I try to drum into customers or potential customers: the fact that they should be very proud of this national heritage," says Carlos. "Then the second educational hurdle is the lack of sweetness in our products. When they try our higher- percentage chocolates, it is another part of the education process to get them to appreciate the stronger flavors and the lack of sweetness." As a result, people who have been to Europe and had higher-percentage chocolates and liked them are Danta's current core audience, but the overall market is growing.

While Carlos targeted his audience directly, El Ceibo in Bolivia, the 100 percent locally owned and run Bolivian tree-to-bar chocolate manufacturer, was completely surprised by its local fine flavor success. After all, Bolivia would seem to have a limited market for fine chocolate. According to Volker Lehmann, who owns a fine flavor farm there, of a population of around ten million, fewer than 5 percent "knows what good or fine chocolate is." That's why, for almost twenty years, none of the two-dozen-plus products El Ceibo produced for the Bolivian market were fine flavor and why their initial foray into fine-flavor production targeted international audiences. Yet El Ceibo, which developed its fine-flavor product line with Chloé Doutre-Roussel in 2008, has had an unexpectedly huge success locally: While 50 percent of its fine flavor sales are in the United States, where they have set up distribution, 30 percent comes from Bolivia, more than double that of Japan, its next largest customer. No one predicted that, especially when it is more expensive than other "premium" chocolates in the market, such as Lindt. As a result, El Ceibo is nurturing the Bolivian market for the future because, as Chloé states, "Shipping and customs charges to Europe are more than five euros per kilo, making it very expensive for them to develop a larger international market for an already very expensive product, so it reduces its competitive potential. If the success can continue in Bolivia, and if consumers remain proud to have a Bolivian product expanding in their own country, this is great for the market."

The same thing is happening throughout South America with recently established manufacturers like Amma creating chocolate specifically designed for a growing Brazilian market. Even more established companies are turning their attention to the local market: Chocolates El Rey in Venezuela devotes considerable time presenting consumers with high-quality chocolate that they are not used to having and is leveraging social media to make it happen. "There's a trend to shift from the crappy stuff that we have been consuming here to the good stuff that we have been presenting," CEO and president Jorge Redmond says. "We launched our San Joaquin chocolate in a 500-gram bar in a special box with super-premium packaging on Twitter. We only had 3,000 units we wanted to test and it was gone in a day. Just from Twitter, all of it locally. That surprised the hell out of us. The lesson we learned from introducing San Joaquin locally has really taught us that we need to be even more active in that area."

At Pacari in Ecuador, Santiago Peralta is already there. In the past, most Ecuadorians could only find high-end Ecuadorian chocolate at the República del Cacao airport stores. Santiago is trying to change all that by introducing his fine flavor bars to the local market. "The Ecuadorians, after years of tasting, tasting, tasting and explaining our chocolate to them, never go back to the milk chocolate or the candy that they used to have," says Santiago. "There are now 20,000 people in Ecuador who eat our chocolate often. That didn't exist two years ago. Before if you asked an Ecuadorian where the best cacao beans on the planet came from, they would say, 'Ecuador.' But if you ask, 'How come?' they'd say, 'I don't know.' They had not experienced the difference. But as soon as they taste it they get it. It is so clear. You don't need a refined palate to understand why this is fruity or this is floral. This connection is clear and is just a matter of education to understand." In fact, Santiago preaches his gospel of fine flavor at the local university, giving talks and organizing tastings for hundreds of students. "But we need a bigger experience," he adds, "more massive, or we will just see it in the fine hotels and airports."

This message resonates with the chocolatiers George Soriano and Julio Fernandez of Sibú Chocolates in Costa Rica. Until recently, Sibú chocolates were available only at hotels and places frequented by tourists. Now a local chain of bookstores, among other places, is selling them. And, much like Carlos Eichenberger in Guatemala, George and Julio have tried to educate as many people as possible on the environmental, social, and economic importance of cacao to build the business. "We thought the market was right in Costa Rica to introduce a Costa Rican-made chocolate. We started out telling people that it is very important that cacao originated from this region. It is an indigenous plant. Coffee is from Ethiopia. Sugarcane comes from India. Bananas come from Malaysia. Cattle are Eurasian. Cacao returns us to our roots. And cacao fields require a certain level of biodiversity, so where you have a cacao plantation, you have lots of life. And in terms of the social component, we are named Sibú, which is the name of the creator god for the Bribri tribe in Costa Rica."

But George and Julio discovered that reaching out to locals and simply telling them about indigenous crops and heritage and biological corridors was not enough. "They thought that was interesting but then so what?" adds George. "We found we could really engage people by telling them we harvested and made the final product and bonbons here—that our fine flavor cacao did not have to be exported to Belgium, Italy, and the United States. That had never been done before. You give them that and have them taste it, and then tell the story that's attached to it—the story that's wrapped up in that piece of chocolate. . . . Today, our chocolate bars are sold mostly to Costa Ricans. A lot of our customers are college-age kids and high school kids who are fascinated by the fact that it is a Costa Rican chocolate made from Costa Rican cacao. You can make fine chocolate in Costa Rica."

Sibú's success has been contagious in Costa Rica. "When we started nobody was doing any good finished chocolates here," says George. "It took us four years and now there are four new chocolate shops and everyone is interested in buying Costa Rican-origin chocolate now. Only the ones who are doing terrible chocolate and charging a fortune are the ones who are doing us a disservice. We hope and want more people to do it right and raise the bar."

This is already happening in the United States, where more than thirty fine flavor chocolate brands have been founded in the United States since 2005. They define the "craft business" spirit Adam Davidson spoke about in his *New York Times Magazine* article "Don't Mock the Artisanal-Pickle Makers." And what

Davidson writes is what every manufacturer hopes for: "When it comes to profit and satisfaction, craft business is showing how American manufacturing can compete in the global economy." In fact, many manufacturers prefer the word "craft" to "artisanal."

"We say we are about craft chocolate, because it's about mastering a hands-on craft," says Rick Mast of Mast Brothers Chocolate. "I'm not asking my guys to come up with silly cool new flavor combinations. I'm asking them to winnow better than anyone has ever winnowed or be the best roaster or keep the temperature level and keep everything nice and clean. Our staff has a pride in this."

Many of these US manufacturers may be small, but they have been driving recent changes for the better in the industry: Change the world—make better chocolate. They pride themselves on direct and transparent trade, paying top dollar for the best beans, speaking out against forced labor, investing in education, and making chocolate that tastes nothing like the multinational mass-market brands. And while most of those manufacturers and chocolatiers in the United States must turn to gourmet stores, specialty shops, the Internet, and supermarkets to sell their products, more and more of them are also building their success in what might be called a very European way: targeting their local communities at markets, events, and their own retail locations and combining that with a factory tour and tasting experience.

For example, Mast Brothers sells 90 percent of its chocolate to people who live within several miles of the store. Rick Mast hopes this success is a sign of a bigger movement in which the majority of retail in the United States becomes local and devoted to small manufacturing again with neighborhood butchers and cheese makers and craftspeople of all kinds available on every corner. "Our mission statement as a company is to provide locally produced craft chocolate," says Rick. "That's it. We don't need to design the packaging or do publicity to make sure that people are educated in Singapore. That is the importance of the local food movement in general."

Whether it is Theo Chocolate in Seattle or TCHO in San Francisco, small manufacturers are opening their doors to packed tours of people eager to learn about flavor, how chocolate is made, and where it comes from. They are even trying to be heroes in their own backyards. Shawn Askinosie, whose factory is located right in the middle of the country, in a lower-income neighborhood in Springfield, Missouri, uses the power of chocolate to educate local high school students in his Chocolate University. The students travel with Shawn to Tanzania as part of the program and then get to taste chocolate made from beans they brought back from their trip—chocolate like they never had before. And . . . with no parents to tell them they might not like it.

"They soaked it up," says Shawn. "I don't know that there has been anything more gratifying for me as a person. This project may very well have an impact on their future. That is a big deal for me. We work with kids in the fourth and fifth grade of our little elementary school, too. We incorporate nutrition into our program just like we do for farmers in the Philippines. It is awfully glamorous to do that in the Philippines, but what about here? What about right in my own backyard? We are in the midst of the program to have them make a product and I'm going to buy it and sell it and all of that money will fund a nutritional program for one hundred kids. But it is all with young people learning at the same time about chocolate."

Of course, tastings, tours, and other programs can only reach so many people and go so far. To complement those tastings, manufacturers worldwide offer a vast amount of information to consumers about their chocolate and fine chocolate in general. For example, manufacturers like Chocolates El Rey in Venezuela breaks down all the flavor components of its chocolates and provides comprehensive detail on flavor, including how consumers might use each chocolate. Industry websites and retailers complement the manufacturers' work and provide glossaries, reference materials, and topical blog posts. John Kehoe of TCHO says one of his company's "biggest contributions to the chocolate world" is how it positions chocolate by flavor with its Flavor Wheel (nutty, earthy, floral, etc.) in order to "help consumers understand that there are a number of flavors out there and that chocolate doesn't just taste like chocolate . . ." Dan Pearson takes pastry chefs who have thirty years' experience but have never seen where chocolate comes from to Marañón Canyon in Peru so they can understand before they taste. The list of similar efforts at education is long and only growing, because in the words of Richard Callebaut of Swiss giant Barry Callebaut: "We put a premium on the importance of education so consumers can make the best decisions."

Moreover, chasing, educating, and satisfying hungry customers is hardly limited to just the growing regions and the established US and international markets. Stories abound about new markets opening up for the future. In India, Barry Callebaut has opened a training school in Mumbai. Reuters reported that

Belgian chocolatier Thierry Detournay has become "Indonesia's Willy Wonka," producing a better chocolate for Indonesia's booming middle class; his Chocolate Monggo now employs fifty people on the production side. And Thierry is not the only Belgian looking past his own chocolate-mad country. The headline of a Financial Times article in November 2011 said it all: "Chocolatiers savour sales to emerging market new rich." According to the article, "If you are a Belgian chocolatier you have to find your 'new rich' elsewhere, particularly—as it turns out—in Russia and the Middle East." Eastern Europe is getting in on the fine flavor craze, too: Chloé Doutre-Roussel is consulting with a bean-to-bar chocolate maker in Giedraičiai, Lithuania and has volunteered help to two others in Budapest, Hungary.

Then there is the biggest potential market: China.

The Chinese market is particularly intriguing to fine flavor manufacturers because the Chinese do not traditionally have a sweet tooth, so fine flavor chocolate has a certain appeal. Angelo Agostoni, president of ICAM in Italy, recalls hosting a delegation of Mainland Chinese customers in Italy a couple of years ago and one person politely tasting the dark chocolate being offered. "He said, 'Wow! This is chocolate I would like to be offered. This is something I love.' It was the first time he had bitten into a dark chocolate. Generally the Chinese are very fond of bitter flavors. Mainstream chocolate brands available in China have been very sweet products, so chocolate in their mind was sweet—which is why he was not so fond of it." Today, the professional chefs ICAM services in China are expressing more and more interest in high-end, single-origin, and organic chocolate. And while this group still represents a small part of the economy, ICAM sees the market continuing to grow as they learn and explore. Barry

Callebaut has recognized this change and has opened a factory an hour from Shanghai as well as a Chocolate Academy.

Shanghai is a more westernized in terms of its tastes, so fine chocolate's presence there is a little less surprising than the success of Polly Lo and Laurier Dubeau's La Place Collection in Beijing, which started as an Internet-only business. "Even the hygienic department didn't know what license to give us because they never heard about a chocolate business in Beijing before," Laurier recalls. "They heard about pastry shops and hotels but not a little chocolate business."

Laurier disputes the idea that the Chinese "don't do chocolate" because they don't eat a lot of sweets. "They do eat a lot of sweet stuff but they don't eat very heavy cream-based and French desserts," Laurier explains. "Chinese people even have their own chocolate, but it tastes like candle wax—even the Chinese, they don't really like it." What the Chinese do have a taste for is gifts, so Polly and Laurier used their business contacts in Beijing to set up shop and sold their chocolates initially as corporate gifts via Internet sales. "We did not think they were ready to just come in and buy a box of chocolates," Laurier says. "They would buy it as a gift. We started with IT companies because IT people were the first ones to go abroad from China. Then we sold to the banks and then oil. China is a big gift culture and they love to give gifts but they don't want to give a gift that you can find anywhere in the supermarket, so that worked for us."

Education hasn't always gone smoothly. La Place's customers often ask if it specializes in Belgian or French chocolate because those are the perceived prestige chocolate countries. Moreover, there was no word for truffles in Chinese, so Polly made up her own expression, *tefu—te* means special in Chinese and *fu* means good fortune or happiness, so the truffles are a "very special happiness" or "very special good fortune." The good fortune seems to be working. According to Laurier, "Now more and more we are finding people saying, 'No, no, no, I don't need a box, this is just for myself, I just want to taste.'"

Because of Laurier and Polly and the efforts of all fine flavor manufacturers around the world, consumers everywhere are becoming more and educated about the origins of cacao and chocolate worldwide. They are learning how flavor varies depending on terroir, postharvest processing, and chocolate making. The results are reflected in recent consumer trends, including what some call the "purist trend," where customers have a preference for a single origin, a bean type, or percentage of cacao. Does that mean that fine flavor manufacturers see fewer flavor trends developing in the future, like maple and coconut sugar sweeteners, or inclusions, like the salts and peppers, fruits and nuts, and everything from pop rocks to coffee to bacon from the past couple of years? Of course not. As Gary Guittard says, "Adding stuff to chocolate is part of the fun of being in chocolate. I think we are pushing the envelope today but then again it all comes down the flavor. If it is something that really tastes good, then it is probably going to have legs. And that is the fun of it all."

Angelo Agostoni of ICAM thinks this is all possible because of the increased level of education among consumers. As Angelo notes, "Consumers have become more educated and even with milk chocolate there is generally higher expectations for quality and flavor. As a result, we have experienced a revived passion and interest for milk chocolate, playing with higher cocoa solids than traditional milk chocolate, or enhancing secondary notes to deliver caramelized milk chocolate or creamy milk chocolate or malted chocolate and even some limited interest in very high quality white chocolate."

"Twenty years ago," Richard Callebaut of Barry Callebaut says, "we did twenty percent dark and eighty percent milk. Today more like fifty-fifty but there is strong interest in milk still." In fact, milk chocolate, particularly higher percentage milk chocolates and caramelized milk profiles, were top of mind for the future for several manufacturers and chocolatiers and are commanding a premium compared to the milk chocolates of yore.

That's certainly true at Valrhona as well, which has had incredible success with the launch of its Caramélia milk chocolate and is pursuing many chocolate varieties. "We are producing a large range of flavors in dark chocolate and milk chocolate but we are also coming out with white chocolate," says Philippe Givre of Valrhona's École du Grand Chocolat in France. "I think our customers will prefer more and more dark chocolate but maybe will rediscover milk and white chocolate with new techniques or textures. In our work at the school, it's so interesting to make 'alliances' of different chocolate flavors with the other ingredients we use in recipes. The more we use different chocolates, the more we learn associations and techniques to sublimate and complement flavors."

Some manufacturers and chocolatiers even admit to *liking* milk chocolate or even—gasp—preferring it. Michal Recchiuti of Recchiuti Confections in the United States says without qualification that he is "really into milk chocolate right now." At Cioccolato Artigianale de Bondt in Italy, one of Paul De Bondt and Cecilia Iacobelli's signature bars is a layered dark and milk chocolate bar with sea salt between the layers, "representing the two of [them] united in one." Others are simply enjoying the challenge of creating excellent milk chocolate bars to attract more consumers into the fine flavor market. For example, Duffy Sheardown of Duffy's Red Star in England will continue to explore "surprising differences" in the milk chocolate bars he manufactures like his 55% Criollo bar made from Venezuelan Ocumare or one made with oak-smoked cocoa nibs.

But whatever it is, Frederick Schilling of Amma and Big Tree Farms says, "People like their chocolate chocolatey and their base chocolate smooth." Which is why manufacturers are focusing on machinery that will help them produce better chocolate and achieve that smooth, melty texture consumers desire more and more in milk *and* dark chocolate. This not so much a change from the past but an aspiration—one that might possibly add another layer to how fine flavor chocolate is marketed with manufacturers talking about their specific roasting or conching times. (Conching is a texture and flavor process done by machine and is similar to kneading.)

Some, like Michael Recchiuti, think this can be overkill. "I think some manufacturers got distracted by the media," he says. "The media has developed a kind of glossary for consumers and manufacturers to follow. The glossary includes kind of origins, beans, percentages, conching . . . these chocolate manufacturers put all this information out there. But I think people don't even know what they're talking about. You ask people, 'What is conching to you?' Why would you ask how long you conch your chocolate for? What does that really mean to you? If it were longer would it be better?"

Anne Weyns at Artisan du Chocolat in England, who processes her bonbons from liquor, seconds Michael. "I think in the end the fine chocolate industry can be a little bit snobby. If people go into a chocolate shop, they should not have to have a PhD in cocoa beans. They go there to buy something that they are going to enjoy eating. That is it. They want something nice. Something they can come back and buy again. It is gone too far into 'this bean has been conched for this long.' People don't really care. They just want a bar or a bonbon that tastes good and is reasonably priced so they can come back and buy it again and again."

But Steve De Vries does see the logic behind this information as he and other manufacturers are always passionately experimenting with those and other techniques and tricks to improve and retain the flavor. "Education is not limited to the consumer," Steve says. "We are all learning and tasting what everybody is doing and it is just exciting to have so many people doing different things. That is going to improve the chocolate."

Of course, not every manufacturer is after the same texture. Taza and Felchlin, for example, continue to have success with their crudo chocolates, which are minimally or unconched. Some think that this minimally or unconched chocolate lacks complexity but Felchlin, which launched its first Grand Cru couverture in 1999, believes quality is not just in the texture but also in the combination of details that must be appreciated through attentive concentration. "Think about a classical concert or painting, the same applies to edible pleasure, we must hone our senses and consciously appreciate the experience," argues Christian Aschwanden, Felchlin's CEO. "This is why Grand Cru must be presented differently. Our effort focuses on this. Many consumers are surprised at their own perceptive abilities. The future existence of Grand Cru relies solely on the selective consumer who can recognize the exclusive value and is prepared to pay the additional cost and is constantly searching for the ultimate flavor adventure."

Conching removes the vinegary acetic acid created by the fermentation process and affects the flavor of the chocolate (for better or worse is a matter of opinion as some flavor is lost but other flavors are developed). For example, Art Pollard left lots of acid in Amano's Dos Rios bar—its strong orange flavor could not be sustained without it. While Taza and Mast Brothers do not conch at all, some manufacturers are finding ways to retain those acidic flavors while still conching. For example, Anne Weyns at Artisan du Chocolat, who works with liquor, not beans, conches before she refines using a machine developed for the paint industry and has cut the conch time from around forty hours to a highly efficient and, in her opinion, taste-preserving half a day or even a few hours. In Peru, Dan Pearson of Marañón Chocolate found a procedure that could remove the acetic acid and retain the fruit acid for flavor in his pure white cacao beans with a conching time of only fourteen hours.

As Dan and Anne note, all this also saves energy—a trend not lost on most manufacturers big and small. "Do you know what would happen if not just me but big companies could take those multimillion-dollar machines and run them sixty-five percent less?" Dan asks. "You could decrease the cost of Big Chocolate. You could give those farmers eight cents on the dollar. And they might even be able to send their kids to college." A change that big would have to come from the top down in the industry to have any huge effect in the future, but Barry Callebaut is getting there in its work. "Some people think you still need to conch for thirty-six or forty-eight hours but that's just not so," says Richard Callebaut. "It depends on the flavor and what you're looking for. Plus the designs of the conches are so much more efficient. These are the kinds of things that have changed but the rest of the process has remained the same. We still go through the same stages that we did thirty or forty years ago. It is just that the way you do it is much more efficient and the machinery is more modern and larger."

In some ways, this is where the big guys have the advantage for the immediate future. That industrial-sized equipment at a Barry Callebaut or Nestlé can, according to Jeffrey Stern, chocolatier at Stern Chocolates in Ecuador, make a chocolate that can be sold at $4 a kilogram. But in Ecuador, he notes even places that process thirty tons a day sell it at close to $6 a kilogram. "That means it's costing them at least four dollars a kilogram to make it. They just can't compete, even with lower labor costs. They just don't have enough capital to develop the economy of scale to produce chocolate on such a massive level and at such a high quality."

Angelo Agostoni seconds that thought, noting that ICAM continually invests in technology not only with the aim of more cost-effective and larger production runs, but also to make higher-quality chocolate even at a bigger company like his. "Consumers might say quality is paired with artisan or little chef, but in chocolate manufacture that is not true," says Angelo. "Which is why we have built the Orsengio plant. Technology costs a lot of money, so yes, by definition if you want to rely on technology you need to have the critical mass that enables you to do so; otherwise you cannot afford a plant like ours. We need to work and grow as well as spend for the technological costs but we think better technology very crucial in this very tight market."

Mott Green of Grenada Chocolate knows this well, which is why he will not stop investing in technology. "Chocolate-making equipment is expensive and for very small scale is very inefficient," he says. "Making it on a small scale sounds very romantic but your equipment is running twenty-four hours a day and it's very expensive to produce the chocolate. There is more labor involved and more energy involved and more space involved." But that's what Mott knows he needs to do to keep making "a better chocolate and a better texture" to meet what consumers expect. And what they are expecting, Mott says, is a chocolate "as creamy as possible, as close to a milk chocolate so it is mild and sweet."

In the end, George Soriano of Sibú sums it up nicely: "The truth is we just keep trying to make the quality better and finer and keep the flavor where we want it to be."

On paper, all these efforts appear to be paying off. Consumers are indeed buying more "premium chocolate"—just consider what is happening in the United States. According to the most recent numbers from Euromonitor International, worldwide chocolate sales are an estimated $102.3 billion with projected increases of 2.1 percent annually through 2016. According to Vreeland & Associates, a confectionery industry market research firm, the US share of that chocolate market grew 6 percent to $19.29 billion in 2011. Vreeland & Associates found "premium chocolate" accounted for $2.7 billion of those sales, or 14 percent, with an expected growth of 10 percent annually. Similarly, the market research group Packaged Facts estimated the US market at $19.5 billion in 2011—a 6.6 percent increase over the previous year. And according to Packaged Facts, "premium chocolate" accounted for even more of those sales: 25 percent or $4.9 billion, up 17 percent since 2006.

What started with Chloé Doutre-Roussel's nightmare seems to be a dream coming true on the strength of small- and medium-size fine flavor manufacturers. So what's the problem? Well, it starts with the fact that "premium chocolate" on paper may not be any more "fine" than the paper it's packaged in.

Actual Reality?

For fine flavor manufacturers, the problem with "premium chocolate" market research numbers gets back to something discussed at the outset of this book: there is no accepted definition of, or universal standard for, "premium chocolate." Is premium chocolate the same as fine flavor chocolate? Is it made from only quality beans, properly harvested, fermented, and dried? Do only the finest ingredients go into manufacturing it? Must it be dark or darker chocolate with high cacao percentages? Is the chocolate organic or certified? Does the wrapper mention or picture the farmers and the beautiful places that grow cacao? Must it come in an expensive box or package with lots of gilded foil, artwork, and images of the mouthwatering treat inside?

The truth is, "premium chocolate" is whatever someone says it is.

Curtis Vreeland of Vreeland & Associates explains that for statistical data mining of sales receipts from International Resources, Inc. or Nielsen, "I use a definition of premium chocolate which has been around for several years: chocolate selling for greater than $8.00 a pound, with adjustments for gift chocolate that includes a lot of markup for packaging. Qualitative factors are: using better quality ingredients, better execution, upscale packaging, etcetera." This seems a fair standard and, depending on what subjective criteria and data are used for premium chocolate sales by other market research firms, might even explain why Vreeland & Associates and Packaged Facts estimates for premium chocolate sales in the United States differ by $2.2 billion.

Yet while premium chocolate as defined in chocolate retailing costs more than $8.00 a pound, fine chocolate starts around $24.00 a pound at retail and can easily exceed $100.00 a pound. This is why numbers like the ones Vreeland & Associates and especially Packaged Facts have for premium chocolate do not match up with those used throughout the fine chocolate industry. That number, offered many times without debate at the July 2011 Fine Chocolate Industry Association meeting, was $5 billion in sales worldwide for fine flavor chocolate. Packaged Facts estimated sales for premium chocolate in the United States *alone* at close to $5 billion. The discrepancy is astounding and can't be completely explained away by saying "fine flavor chocolate" is only a percentage of the chocolates considered "premium chocolate."

Why not? As many people in the fine flavor industry say, 80, 85, and up to 90 percent of what you read on the average chocolate package is "marketing" or "lies" or "propaganda." Remember: "premium chocolate" and "fine flavor chocolate" are whatever someone says they are. And there is no agency to expose the truth, no referee, no single voice for education. There is no Cup of Excellence program like in the coffee industry—

something Volker Lehmann in Bolivia desperately wants—that could help motivate growers to invest in quality and usher in more transparency, quality, and choices for customers. The Fine Chocolate Industry Association's Heirloom Cacao Preservation Initiative may help change that, but until that day, as Dan Pearson notes, the United States will have only one standard for chocolate: "In the United States, to be called chocolate, you only have to have fifteen percent liquor. It can be eighty-five percent sugar. All these people, including myself, who grow up saying we are addicted to chocolate? We are addicted to sugar."

While the minimum standard varies country to country, it does not change the fact that for the most part, companies can call chocolate whatever they want beyond that minimum standard. Sure, some information is verifiable, like nutritional information and percentage of cacao, but, as discussed in Part One, one 70% bar could have no added cocoa butter and another could have 20%, making it a lot less powerful flavor-wise. And even then percentage is no indicator of quality per se. Go into any store that sells a variety of chocolate and you will see a lot of high percentage chocolate marked "premium" and "fine." The premium chocolate is three-and-a-half ounces for $2.99 and a Hershey bar of the same size is ninety-nine cents. But what is the difference? Not much, many argue, and they worry consumers are being fooled into thinking otherwise, which is why Angelo Agostoni, president of ICAM in Italy, could be speaking for many in the business when he says, "The only danger I see in the future of fine chocolate? There is too much marketing and hype."

Art Pollard of Amano Chocolate would concur. "You have an immense number of companies who purport to have some of the finest chocolate in the world and you taste it and it is quite clearly made with average quality African beans," he says. "The importance of truth in marketing is paramount." Because it is marketing that is driving many of the decisions consumers make. As Christian Aschwanden, CEO of Felchlin says, "A poor quality cacao can be disguised by other dominant aromas. Marketing is sometimes more important than the flavor and quality of a product because the consumer is not in a position or doesn't take the time to distinguish quality chocolate." That's one of the main reasons Joe Whinney of Theo Chocolate says: "I can't really think of one major brand that hasn't made some sort of enhanced claim. So that's also why I think transparency is important. Products have to be three-dimensional in terms of the product quality, its price, and value proposition, and the impact that it is having on the community and the rest of the world. That's where the future is."

If that is where the future is, it cannot happen too soon for many fine flavor manufacturers. They worry about a backlash founded on their efforts to educate consumers and have them pay more for fine flavor chocolate that results in a lower demand for chocolate overall, not a higher demand for their fine flavor bars. Not that there isn't a place for affordable step-up brands. There is a wide-open space between Cadbury Dairy Milk and a bar made from expertly harvested and processed 100 percent, single-origin, wild Bolivian cacao. Good chocolate for a good price: many manufacturers see this as a great way to grow the market from the bottom up, much as wine did in the last century.

In fact, while people in chocolate make analogies to everything from studio art to craft beer to olive oil to explain their hopes for the future of fine flavor chocolate, it is wine that is most often invoked. Wine drinkers worldwide have long moved beyond France, Italy, and California to appreciate the diversity of wine-producing regions and estates. Through years of tasting, education, and rating systems from publications, experts, and retail stores, wine drinkers have not only learned their favorite wine types and producers but also can ask for different types of wines by name, grape, or region, often in a range of prices. They also know that most wines will vary in flavor from year to year. Fine flavor manufacturers hope that the same thing will happen with their chocolate as consumers learn to slow down, taste, and appreciate chocolate.

"I think it is a lot like the education for drinking wine," says Michelle Morgan, who has certainly seen Australian wines make big names for themselves at all price points. "Chocolate changes not just in terms of the batch that's being made, but in terms of the crop changing each year as well. Consumers should stop expecting this monotone sameness like the big guys are trying to achieve. This is a crop that changes and has its own beauty in it. Enjoy it now because it may never taste like that again. Really, really value it. It is getting people to understand about that part and just please taste it. Just step back and try to forget everything else and taste it and see what you think."

This gets back again to the battle cry for education as fine flavor manufacturers chase customers around the globe: "Taste! Treat chocolate as a food and understand the flavor and the story behind what you are eating!" People like Michael Recchiuti say this to the people who are just beginning to explore the world of fine flavor chocolate as well as those people who are—for whatever reason, educated chocoholics or not—members of what he calls the "Seventy Percent Club."

"I don't know why people feel that number is the number or anything above that is the number," says Michael, who has removed cocoa percentages from all his chocolate bars and bonbons. "It's taste. It's flavor. It is like people who salt their food before they taste it. That really pisses me off. Just try it, taste it."

Thomas Haas of Thomas Haas Chocolates in Canada seconds Michael Recchiuti when he calls everyone in the chocolate world "victims of marketing" when it comes down to information like percentages and origins—even when it comes to his own products. Like Michael with his percentages, Thomas is planning to take down references to origins on his website and in his stores. "Everybody uses it as a marketing tool," he says. "We say this comes from an origin here or this comes from an origin there or this comes from a single plantation. In my opinion that doesn't tell you anything. What I would be looking for, as a customer, is a chocolate that tastes great. What plantation it comes from doesn't tell me anything. At the end of the day, as much as we do tastings and as much as we debate and as many suppliers we have come in to bring us samples, it comes down to one thing: Does it taste great or not? Anything else is more marketing than reality."

Steve De Vries could not agree more. Now and in the future he just wants people to trust their mouths and taste different chocolates, *then* make the decision as to whether it's worth it. But, he says, "They won't. Ten percent will and ninety percent will just follow them. So you have to kind of go after that ten percent and then the other people will start following. But that can be madness if tastes change quickly." It can also be madness for a consumer confronted with the sheer number of choices on the market. "The explosion of small manufacturers means a dizzying array of fine flavor options from around the world, each with its own taste and story," Steve adds. "And that can just be confusing."

This confusion is only compounded by the vast amount of information that these manufacturers offer about these origins and their chocolate—for both education and marketing purposes—making it harder and harder for consumers to choose and distinguish between brands as they explore. But no matter how much information manufacturers and chocolatiers put on their packaging, websites, and marketing materials about their chocolate—its origin, processing, ingredients, percentages, the people behind it all—they are essentially talking about the same thing. This has led to an even bigger quest for differentiation through inclusions and even the look of the chocolate itself.

"A lot of these chocolate makers—I taste the chocolate and they are so flavor- heavy that you are not getting to experience the chocolate," says Michael Recchiuti. "They're interested in the crazy combinations and different colors and shiny chocolate but they are not really interested in the taste. I'm hoping that the trend really kind of shifts into taste. I am preaching that when I talk to people getting into the business. You do not need to be the next trend or in the Seventy Percent Club."

Off the record, some in the industry go even further than Michael. They call for a pullback—an actual reduction in the number of products, if not manufacturers—in the name of taste and stability. Whether this pullback will happen or is even necessary remains to be seen; and if it does happen, it will be driven by many factors beyond taste such as education, marketing, price, and supply of cacao. But a need for stability in flavor is one of the reasons that many fine flavor manufacturers, like winemakers, have turned to blends to satisfy their customers' tastes now and for the future.

Blending In

In the world of fine flavor chocolate, sameness has become a benefit and a necessity. Not sameness as it applies to multinational mass-market chocolate companies seeking out low-quality beans to keep costs down but sameness in terms of stability and consistency. According to some, blending can even make for a better tasting chocolate; a blended bar using pure Ancient Criollo can actually enhance that Criollo's flavor profile. As long as the packaging reflects the actual content and the processing is expertly handled, most manufacturers have no problem with that blending. In fact, blending could actually help consumers reconcile their expectations for sameness with a new appreciation for quality and flavor. Chloé Doutre-Roussel, for one, freely admits that she created her own blends as a "reaction against the brainwash of single origin and single plantation."

And since cacao, like wine grapes—even the bulk varieties—is subject to agricultural variation, sameness is often impossible unless a manufacturer can occasionally adjust for those variations with a lower quality bean to achieve uniformity. Steve De Vries recounts how he once got some incredible beans from Chuao and made some single-origin chocolate out of it. Later, he tasted a larger manufacturer's Chuao. He could taste similarities but it was not quite as robust as his. He called the manufacturer and asked, "'What do you do when you get a bean that is head-and-shoulders above the normal bean from that place? You have to step on it, don't you?' And they told me,

'Yeah, if it is too good we have to push it down a little bit because if we put something out that is better than what we think we are able to do next year, we will have to deal with complaints next year.' Could you imagine a wine vintner tasting the wine and saying that it is good but a little too good so let's pour some water in there and step on that a little bit?"

Probably not, but it is clear that blending is a growing trend in the fine flavor industry, especially for manufacturers providing couverture to chefs and chocolatiers. Gary Guittard puts it all in perspective when he says, "Chocolate on a whole basis is blended beans. It is blended beans, and chocolate makers want to be very protective and secretive about those plans because some blends are very synergistic and the whole is greater than the sum of the parts."

At Guittard, Gary's first experience of this trend was with the company's Harmony blend. "We tasted a whole bunch of different blends and all of a sudden we tasted this blend and we went, 'Wow!' It just stood out over and above all of the other blends that we had put together," he says. But even with a fine flavor blend, Gary notes, replication is not always possible: "We stopped making Harmony for a while because we couldn't get some of the beans for the blend. But that gets to the reason people use beans in blends: sometimes these flavor beans aren't always available or you can't get them in the quality you need or perfectly fermented, for example, but then you can substitute something else without drastically changing the flavor of the blend."

That's exactly the reason Pierre Hermé has a close and exclusive relationship with Valrhona. While he says he will taste any chocolate—blended or single origin—sent to him, he remains exclusively devoted to his Valrhona blends. Valrhona delivers what he expects and they explore new directions together. "I am always curious and happy to discover new products and I have tasted many wonderful products," says Pierre. "But I only work with Valrhona, since we have a unique relationship that gives me a lot of freedom for creativity. We collaborate, sometimes for years, on developing together a recipe that fits my needs. We work on the origin, the fermentation, the roasting, the conching—they have a strong know-how in their own field, I have one in mine, and we collaborate using our respective know-hows to get my couverture."

For fine flavor chocolatiers and chefs like Hermé, consistency from a manufacturer is essential. Even an artisan like Michael Recchiuti in the United States, who has a similar relationship as Pierre Hermé with Valrhona (though not exclusive), and who embraces flavor nuances and the differences from year to year, still needs a consistent blended product for his signature bonbons. Jorge Redmond of Chocolates El Rey in Venezuela, which makes blends for Michael, likens the blending process to what is done in perfume, wine making, and painting, and touts it as a growing part of El Rey's business. "When we first started, we were making three types of formulations. An extra bitter, bitter, and milk," says Jorge. "Then we found that Michael Recchiuti wanted a more fluid chocolate for his needs. As we learned this and other things from our customers, we became more professional in what we offered." El Rey now offers six formulations and has not ruled out offering more or creating additional custom blends for its biggest customers.

Simply put, small manufacturers may tout the fact that their single-origin bars will taste different every year but chefs and chocolatiers around the world need and expect uniformity for their products. As Mark Adriaenssens, head of R&D in the Americas for Barry Callebaut, notes, chefs and chocolatiers need a consistent couverture for their praline, ganache, and fillings: "They are playing with other flavors. They may feature single origins in pure bonbons or bars. But the power of that single origin is lost in the execution elsewhere and so is the benefit of using it at all."

Chef Bart Van Cauwenberghe, who owns De Zwarte Vos in Deinze, Belgium, and serves as a Belgian Chocolate Ambassador for Barry Callebaut, actually goes one step further. He often gets asked for his opinion on new blend or flavor of chocolate by Barry Callebaut and has one answer: "I always say, 'Please stop creating new flavors because the people are not used to that other flavor and you have already produced a new one. Why?' I could have eighty different kinds of chocolate in my shop. I prefer to have twenty. We use blends and, sure, we could also use single-origin—most of the people who come into my shop, they want to taste something new. But I don't have the time to rebalance the flavors and create new bonbons for all the different chocolates."

The possibilities in blending are not lost on small manufacturers, either. David Castellan at Soma constantly uses different origins in his bars and thinks it is really interesting to do what he calls "intelligent blending." He even names the blends so that consumers can identify them beyond the origin. His award-winning Three Amigos combines Chuao, Alto Beni, and Dominican La Red. "A lot of people wouldn't take a Chuao and make a blend out of it. Most people just make a pure bar," David says. "So we mess around that way."

Of course, this messing around is not just reserved for blends. As long as they know the approximate flavor profile and the quality of the processing, several chocolatiers and bakers working with chocolate plan on incorporating more origin beans and chocolate into their future work despite the variations. Roger and Andrea von Rotz own the five von Rotz Patisseries in Switzerland and are the exclusive Swiss confectioners for Marañón Chocolate's Fortunato No. 4 from Peru; they plan on focusing collections on it even more in the future. "I am aware that due to seasonal and weather inconsistencies there can be only a limited quantity. But they guarantee that the quality is ensured," Roger says. Since Gary Guittard approached Michael Recchiuti with single- origin chocolates that Michael thought were fantastic, Michael has every intention of exploring the flavors further on an educational level when he opens up his Chocolate Lab café. "When I told Valrhona I decided to go with the São Tomé single origin for my milk chocolate, they said it's harder to control that on a consistency level as far as flavor profiles," Michael recalls. "And I said, 'I don't really care.' If it changes, I know it's going to be good. It is just going to be different."

This all adds to the diversity of flavor for manufacturers and chocolatiers to explore. "The point we are at now is we are learning what is possible, like painters," says Steve De Vries. "We are still learning what's possible with the primary colors, but I think blending is really going to be something, and some people are starting to do it now because you can make some complexity of flavor that is really interesting."

So what's the problem?

According to people like Chloé Doutre-Roussel, what Steve De Vries finds interesting is not resulting in more gratification for her as a consumer—she is finding less, not more, pleasure, despite the fine flavor revolution she has witnessed. "Globally, as a consumer, the world of fine chocolate is much more interesting and much bigger, but I don't get the pleasure that I had three to five years ago," she says. "Despite having more brands on the market, I'm not experiencing more pleasure. There are great new chocolate brands, but too many other newcomers are in a hurry to launch their products with no use of their brains, no research, no focus on quality—just marketing. This means that we, as consumers, have more choices but a harder time finding fine chocolate. And I'm not willing to spend my money on it, at least not at that price."

That lack of a value proposition is also where fine flavor chocolate and wines remain separated: Today, people will pay hundreds of dollars for a bottle of wine but many of those same people flinch at $10 for a bar of chocolate. They might be willing to pay $100 a pound for the best bonbons because they reflect a level of artistic skill and vision, but consumers do not yet understand that an incredible amount of skill, vision, and work goes into producing fine quality chocolate itself. And unlike wine, chocolate may be something that people will always really find difficult to pay more money for because they see it as a childhood indulgence—a "treat."

Frederick Schilling, who founded Dagoba in 2001 and sold it to Hershey's in 2006, is similarly concerned. "I knew what was coming down the pipeline with all these small manufacturers. I didn't know how these guys were going to survive. I say this not in a vicious way. I know how cash-intensive any company is, but especially chocolate. All these little companies—they all ask my opinion, and I say this is a lot of money, a lot of work." Simply put, at some point, in addition to all the good that they do, businesses need to make money or they cannot last. Passion only gets them so far—and marketing may be just as important as quality when it comes to generating sales from one-on-one tastings to social media and beyond.

Mott Green agrees about the marketing concerns. Echoing the sentiments of many small manufacturers devoting so much time and energy to doing well by doing good, he knows his chocolate is good enough to command a higher price. But because of a lack of marketing and increased competition—and because he can sell only so many bars at the chocolate specialty shops where the retail price can be a much fairer $6 or $7—he needs better marketing to compete with the big multinational brands and the older established names. "Very few people are willing to spend that much money on a chocolate bar," he says. "Maybe the ones who are going to specialty shops, but not in higher-end supermarkets like Waitrose in the United Kingdom and Whole Foods in the United States. We need the high-end grocery market. But we basically find that over four dollars in the US and three pounds in the UK they just don't really sell." This is especially true in fine flavor-loving countries like Spain that are faced with deep economic problems. "Spain was the gateway to cocoa in Europe. There is a long tradition of quality chocolate, especially in the north where the climate is cooler, and historically, proximity to France has given us a more gourmet vision of the product," says Ramon Morató, master confectioner and director of Barry Callebaut's Spanish Chocolate Academy. "The current challenge in these times of economic crisis is to know how to maintain the quality level and prevent consumers from opting for lower-priced products."

Still, many in the industry remain cautiously optimistic. "Whenever I get concerned about the chocolate market being too crowded, I turn to the wine aisle," says John Kehoe at TCHO. "There are hundreds of labels. There are bottles for forty and eighty dollars and bottles for five dollars and everywhere in between. I see the strength of the chocolate segment in the middle of that range: the two-dollar-and-fifty-cent bar to maybe four dollars and there is a lot there. But exceptional chocolate takes exceptional effort throughout and that creates a more expensive bar. It has to be. But I just don't know if the consumer market is there yet. I mean you're going to have six-, eight-, ten-, and twelve-dollar chocolate bars. Can these be demanded? There are challenges, but I think we can if we make better chocolate, and if consumers dive deep and embrace quality and flavor and support it by spreading the word and telling people when they like a bar that this one is truly fantastic and is really worth it."

Shawn Askinosie of Askinosie Chocolate agrees. In fact, in his graduation speech at Missouri State University, he refused to see a no-win scenario for the future and invoked *Star Trek* and the lesson of Captain Kirk defeating the Starfleet Academy leadership test, the Kobayashi Maru. He urged students to do what he and Kirk did: redefine the problem and instead of asking how much money can we make or are we making, ask a different question, namely how can we make a difference? A successful former lawyer who came to chocolate as a second career, Shawn freely admitted in that speech that he thought he would be in profit or high-profit territory within a few years, but that as a chocolate maker he makes one-tenth what he made as a lawyer and has been forced to scale back. Now he asks a different question for the future: How can we make enough to do what we do at the highest quality, turn a profit, and most important, make a difference? "Is it measurable? Yes. Is it a standard definition of business? No. I redefined the question. Yes, I need to pay everyone including myself but I also started Chocolate University. I took thirteen lower-income high school students to Tanzania. One texted his mom and said, 'This is the best day of my life.' That is success because we redefined the question."

But redefining the question still comes down to redefining the price consumers will pay, and Shawn hopes that is possible. "I think some day chocolate bars will be twenty dollars," Shawn tells us. "I want to see the farmers get more for their beans and I want to see the consuming public ready and willing to pay for that premium product. There aren't many premium anythings that you can buy for eight dollars. You're not going to get the best bottle of wine. You're not going to get the best bottle of oil or anything. I don't think you can buy the best bar of soap for eight dollars. But we have to produce the best chocolate that we can and the highest quality possible or we won't survive and be permitted to do these other good things that we do."

In the end, of course, it is the customers who pay the rent and keep fine flavor manufacturers from, as the French say, "putting the keys under the door." As Christian Aschwanden of Felchlin, says, "Taste is personal and individual and in the end the consumer dictates what he wants."

And what a growing number of those consumers want is form before flavor.

Form Versus Flavor: The Future and Functionality

Late in the movie *The Princess Bride*, the hero, Westley, is rendered "mostly dead" by the evil Prince Humperdinck. Westley's friends find him and drag his limp body to Miracle Max, who agrees to create a miracle pill to revive Westley so he can destroy Humperdinck and save Princess Buttercup, Westley's true love. A pill that powerful clearly requires some extra magic. Thus, as Miracle Max and his wife finish their work, they carefully coat their pill in chocolate, turning it into a bonbon.

"That's a miracle pill?" Westley's friend asks, staring in disbelief.

"The chocolate coating makes it go down easier," Miracle Max's wife explains.

Well, of course it does. Miracle Max and billions of us know chocolate makes anything go down easier and "functional" chocolate—sold for its health benefits, to deliver the benefits of other ingredients, and modified to address dietary restrictions—shows every sign of remaining a growth area in the industry as a whole. In a more health-conscious world, knowing that chocolate possesses certain health benefits only increases its "good for you" allure. In terms of unmodified chocolate being good for you, the future seems bright for chocoholics: every few months another study appears touting cardiovascular, antiaging, mood-enhancing, and other healthful benefits from eating chocolate, particularly dark chocolate. A 2012 study published in the *Archives of Internal Medicine* even found that despite chocolate's fat content and added sugar, people who eat chocolate frequently have lower body mass indexes, possibly because it boosts their metabolisms.

Barry Callebaut definitely sees health facts being emphasized more and more both by manufacturers and consumers in the future and will continue to explore these options from a flavor perspective. "We have an innovation section where we try to think about what new things can we do with chocolate," says Richard Callebaut. "We know chocolate has 400 flavor compounds. What can we do with them? We have our dark and light ACTICOA chocolate, where we retained up to eighty percent of the flavonols. That gives you many more antioxidants in your end chocolate." These days, Barry Callebaut is hardly alone. For example, Roquette America launched a chocolate bar enhanced with a pea protein that boosts the nutrition without "changing the chocolate's texture, taste, or processing conditions." Meanwhile, TCHO takes a lighter approach, listing the myriad health benefits attributed to chocolate on its website and linking them to its TCHO-A-Day: a one-, two-, and three-month supply of eight-gram "doses" of chocolate packaged in what it calls a "drug" dispenser bottle similar to a vitamin bottle.

The fine flavor world has adjusted to the health needs of consumers in other ways, too. A hike in demand for lactose-free products prompted Barry Callebaut to create a 100 percent dairy-free alternative to milk chocolate in 2010. Across the board, sugar-free options are proliferating, as more shelf space is being claimed by high-cocoa-content dark chocolate to meet the chocolate needs of diabetics and those who eschew refined sugars. In general terms, since most dark chocolate is vegan, has no added sodium, and is gluten-free, it also appeals to the millions of people who need or prefer such products. An increasing number of fine flavor manufacturers like Bonnat,

Domori, and Pacari, and chocolatiers from Laurier Dubeau at La Place Collection in Beijing to Anne Weyns at Artisan du Chocolat in England to Patrick Roger in Paris also offer sugar-free (with the sugar replaced by a sugar substitute) or 100% bars (with no added sweetener) or bonbons to satisfy not just dieters and diabetics but also those who prefer their chocolate that way.

Of course, while chocolate certainly revives many of us when we feel mostly dead, no one has found a way to make a pill as powerful as Miracle Max's. But people are trying; chocolate is being used more and more to deliver benefits consumers seek from simple protein to calcium to probiotics/prebiotics. Consumer interest in this "functional chocolate," which is enhanced with vitamins, compounds, and other ingredients or chemicals, is strong but remains off the radar for most in the fine flavor world for one reason: flavor.

"If you really want to know what everybody wants, it's the same flavor profile with less calories. And we are studying that," says Richard Callebaut. "But it is essential that it tastes good. People will not sacrifice taste to have less calories but if we could have less calories with the same good taste than that would be huge." Angelo Agostoni, president of ICAM, would concur. He finds functional chocolate interesting but as long as it is "designed and purchased only because it's good for you, it's a short-lived product."

So is it possible to make a chocolate that is highly functional and delicious? "That is the challenge and the question," says Angelo. "But chocolate is mainly pleasure. If we want something good for us, take a pill or some medicine but not good chocolate." On the other side of the world from ICAM in Australia, Michelle Morgan of Zokoko echoes Angelo's thoughts.

She says functional chocolate was on everyone's radar a few years ago in Australia but she did not—and still does not—see it fitting with her brand. "If I want to add natural flavorings to a bonbon, I can," she explains. "I'm not doing anything for health reasons, and I will never put on the box some kind of health claim. I am a believer that if I want more antioxidants I should be eating more fruits and veggies. Just have a healthier diet and then indulge in a piece of amazing chocolate."

Roger von Rotz of von Rotz Patisseries in Switzerland calls functional chocolate and all experiments on chocolate to make it anything but great tasting "pseudo products that do not have anything to do with chocolate." That's exactly why Gary Guittard doesn't see functional chocolate going very far in the flavor world, though he has had inquiries. "As far as chocolate with probiotics or vitamins, that's trying to put chocolate into an arena where I just don't know that most fine flavor people are willing to take it. We have had some customers who wanted to make a calcium-type product for women. But to deliver that in a large-scale way, you have distribution issues—plus, chocolate melts. There are far better ways to take supplements than in chocolate. Unless that particular supplement enhances the flavor and makes chocolate taste better, just like with ginger, then I don't think it is going to work."

"But," Gary pauses and smiles, "maybe if you put Viagra in it or something, it would work." Whether Pfizer is working on that or not we could not confirm.

Pierre Hermé sums up the feelings of most people in the fine flavor industry when it comes to functional chocolate: Pleasure must be the point of savoring fine flavor chocolate. "When I create a product, my only focus is pleasure, my pleasure, the pleasure of my customers." He says, "If by accident, some of the ingredients in my recipe are 'functional,' then it's great but I do not integrate them at all in the creative process."

The problem is that with functional chocolate, form is the first criteria of choice, not flavor, pleasure, the creative process, or even quality. And perhaps no form of chocolate stirs more emotions than raw chocolate.

The Raw Deal

Raw chocolate has only been widely available for a few years and remains a culinary niche but a growing one for both practitioners of raw food diets and consumers seeking more nutritious versions of the food they love.

Like all raw food, raw chocolate has its supporters and critics, but perhaps because it is chocolate, passions on both sides run hot. Raw food eaters are thrilled to have chocolate to enjoy and tout the holistic benefits and high nutrient and antioxidant value of raw cacao. They see it as more nutritional than traditional chocolate and believe that the rawness brings out the natural qualities of the fruit. Meanwhile, critics dismiss it as a fad or, as Patrick Roger says, a "California thing." (Raw food is most commonly associated with the western United States, but there are now more than a dozen raw chocolate companies on the east coast of the United States alone.) Some question whether raw chocolate is even raw, calling it a "lie" based on a belief that the beans have been fermented and dried and are merely unroasted and unprocessed. (In all fairness, the raw chocolate manufacturers we spoke to also believe that some raw chocolate is indeed not raw and feel just as annoyed as the critics about this packaging deception. Moreover, a small number of manufacturers are being transparent about this, producing chocolate that is labeled unroasted or "virgin" but not raw.) But what the critics mostly question is raw chocolate's safety.

Raw food proponents believe cooking food at high temperatures diminishes its nutrients, and basic science is on their side: heating fresh food can destroy some nutritional content. That's why the maximum heat allowed for cooking by any branch of the raw food movement is 118 degrees Fahrenheit, well below the higher heats required for conventional drying, roasting, and conching cocoa beans. Traditional fermentation alone can reach heats of 125 degrees Fahrenheit, so raw chocolate must not be fermented and processed conventionally, which is what has its critics concerned.

For their part, raw supporters do not dispute that unconventionally fermented and unroasted cacao can host bacteria, fungi, and yeasts indigenous to the tropics, and they do see traditional fermentation and drying as natural processes but ones that nonetheless reduce nutritional value. And ultimately, raw manufacturers also claim that general safety concerns are unfounded; their kill steps that compensate for lower-temperature processing eliminate the same bacteria, fungi, and yeasts as the conventional methods. In truth, there are no regulations in place to prove otherwise; no government requires any chocolate manufacturer to follow specific chocolate safety standards. But still raw chocolate raises concerns for the future, particularly among the most established chocolate manufacturers. For instance, Richard Callebaut of Barry Callebaut would not be surprised if the FDA in the United States issued some kind of warning about salmonella in raw chocolate in the near future as it did years ago with raw or undercooked eggs.

In fact, off the record, what the most vehement critics of raw chocolate worry most about is that raw chocolate may actually kill someone and, in turn, discredit the industry. If an outbreak of a food-borne illness related to raw chocolate were to happen, the reaction against chocolate in general could cost farmers and manufacturers their livelihoods. Critics point to the E. coli outbreak traced to organic spinach and more recently the listeria outbreak from cantaloupe and the impact they had on those industries. Illnesses have so far been rare in chocolate, but when a Hershey's plant in Canada was shut down in 2006 because of salmonella—an outbreak eventually tied to soy lecithin, not the cocoa products—there were reverberations throughout the industry. An outbreak at a small bean-to-bar manufacturer—raw or otherwise—could, as David Castellan of Soma states, "ruin it for everyone."

To raw manufacturers and their proponents, this is pure hyperbole: Any food, and thus any chocolate, not just raw chocolate, can have safety issues. But unless any steps are taken, concerns about safety with raw chocolate will remain a war of words with raw chocolate manufacturers and consumers standing by their processes and critics continuing to question whether raw chocolate is safe at all—especially without the widespread testing that most traditional chocolate companies have in place. A company as big as Mars may have four levels of validation to ensure that no bacteria survives the processing, and even many of the smallest fine flavor manufacturers send their fully heated cooked chocolate out for safety analysis. Many raw manufacturers do, too, but not enough to allay the concerns of some like Gary Guittard, who notes, "Chocolate comes from the jungle. It is left out in the open. It is susceptible to birds. Raw is taking a big chance in regard to some of those issues. Some people are using the whole bean, shell and all, and I think there are real dangers in doing that. And I think you might have some rheology issues with raw chocolate because there's a lot of moisture in it. Suffice it to say we have some real issues with raw. There are things that can be done that can mitigate some of this of course, but I am not sure they are being done."

The reality is there is only one way for raw chocolate to be guaranteed safe from the ground up: complete transparency of the entire process from tree to shipping to bar, and 100 percent quality control throughout that chain—something that is difficult even for traditional direct-trade manufacturers at origin. That's why fine flavor manufacturers like Art Pollard of Amano will not manufacture raw chocolate. Simply put, they do not believe it can be safely manufactured. "I think there is a spot for raw chocolate," says Art. "But I think it's best left to the homemade people."

Joe Whinney of Theo Chocolate concurs but sees value on the nutritional side. He loves raw food and thinks raw chocolate is interesting and can have merit: "My grandmother would boil cabbage until it was white—I know that it is true the more that you cook something the more nutritional value is lost. So I think what consumers are asking for is something more nutritious." Still, Theo Chocolate in Seattle, Washington—a seemingly ideal market to launch a raw bar—has not gone in that direction. "I'm skeptical of the set standard that you don't go over 120 degrees when it comes to chocolate," Joe continues. "That's really hard to do in processing a good chocolate regardless of the nutritional value you lose. What I would like to see is a little bit more intelligence and a little less blunt label of 'raw.' Let's look at what the nutritional value is and how we measure that against the processes we use. I think this may get more refined as the raw chocolate movement gets larger."

But even as safety concerns are being addressed, the biggest problem with raw chocolate for many fine flavor manufacturers is that they just do not like the taste of it. "If we could make a raw chocolate that tasted fantastic? I'm open to that," says John Kehoe of TCHO. Michael Recchiuti recalls tasting a line of raw chocolates and thinking some of them were kind of cool. But, he adds, they had "a lot of tannins and a lot of acid and they were really funky. So I asked the manufacturer, 'How can you make that part better?' And he said. 'Can you do that with raw chocolate?' I said, 'I don't know.'"

That unknown is why Gary Guittard echoes John and Michael when he says, "For me it comes down to flavor and is the flavor there? If it tasted really, really good I would probably be more concerned about it for our business. You can intellectualize all you want about nutrition but where the rubber meets the road is: how does it taste?" David Castellan of Soma agrees: "People see 'raw chocolate' and they think it is better for them. There's a whole world of people who are interested in that sort of thing. All the people who like raw chocolate go to farmers' markets and health stores to get it and think it is really good and that is how chocolate is supposed to taste. But it's this terrible, terrible-tasting product with no roasting involved. Meanwhile, here I am roasting and spending all sorts of money on roasting equipment."

Of course, David admits, the people buying raw chocolate are not necessarily the people who would want to buy a bar of Chuao. "They have a different idea of what they want to eat," he says. "They aren't interested in chocolate in the same way that I am. At the same time, it would be better if people understood the process a little bit more. I'd like to explain to people who like raw chocolate what happens during the roasting in the complexities and nuances that they're missing." In fact, one of Soma's most popular items, called Old School, often gets confused with raw chocolate. "We take the nibs and we just grind them in the melanger with sugar; that makes a paste that is quite crumbly and not refined it all. There is no refining or conching, but we do roast it. So some people look at that and think it must compare to raw but it is really just an unrefined chocolate."

Yes, raw chocolate is even more unrefined than that "Old School" creation, but an increasing number of raw manufacturers are working hard to challenge and change these absolute perceptions of raw. Vanessa Barg of Gnosis Chocolate in New York City knows raw chocolate is and can be so much more. She appreciates that many people in the fine flavor chocolate business looking at raw chocolate from the outside in still think like David and the others, but to her and a growing number of manufacturers what is really "old school" is what and where raw chocolate was five years ago and thinking nothing has changed.

"The biggest misconception of raw chocolate is that things are the same as they were when it first started. The word 'raw' calls to mind something that is not elegant—something that is just thrown together. Maybe because a few years ago that was true," Vanessa says. "But raw has evolved just like the industry as a whole has evolved. It has tapped into this entrepreneurial spirit that has drawn so many wonderful people to chocolate. Raw is a play on process and it is exciting to play with flavor and the different possibilities within fermentation and roasting. It may be a simple product by name but the work that goes into making great raw chocolate is just as complex as making any great chocolate. Raw may be unrefined but it can still be elegant—and delicious."

When Gnosis started in 2008, this declaration would have been impossible; it was the only raw chocolate company Vanessa knew about and she wasn't far from wrong. That did not last long—raw manufacturers and raw chocolate products started to appear widely within two years—yet few raw manufacturers were taking the path she chose, namely to explore the broader world of chocolate and flavor in order to produce a flavorful raw chocolate. With a background in health and nutrition, Vanessa believed in the benefits of raw foods, but recalling her passion for the culinary arts, she found the idea of food without flavor appalling. Yet there was not much to work with when she launched four years ago: There were only two sources of raw liquor and few exploring the possibilities of doing more with flavor. So she did. Within a year of opening Gnosis, she had started attending Fine Chocolate Industry Association meetings (eventually becoming a Founding Circle member of its Heirloom Cacao Preservation Initiative). She even went to work with François Pralus in France, who, despite not quite understanding what raw was all about, welcomed her, improved her appreciation of fine flavor chocolate as a whole, and helped her develop a raw liquor from Madagascar beans.

No one else was doing all that in 2009—no one else in raw chocolate, at least. It was all about raw first then, and Vanessa acknowledges that this continues to contribute to the lingering perception of raw as flavorless. "I'll never forget a few years ago when a company called and wanted to buy some liquor from me," Vanessa recalls. "I listed all the liquors I had including the Madagascar from Pralus. I asked him what flavor he wanted and he said he did not care what it tasted like or if it was truly raw. He just wanted to make raw chocolate." Sigh.

Today, flavor is a growing part of the raw equation—a big and quick step for a form of chocolate that has been around for less than a decade. Gnosis is hardly the only raw chocolate member of the Fine Chocolate Industry Association or the only raw chocolate manufacturer being transparent about its processes and exercising the same strong, if not stronger, safety protocols as traditional chocolate manufacturers. Like Gnosis, these raw manufacturers tout not only the health and nutritional benefits of their chocolate but also the same things about their chocolate that fine flavor manufacturers do: origin, direct sourcing, quality beans, integrity, and more.

Gnosis and others may face an uphill battle with some critics, but several fine flavor manufacturers have responded to Vanessa Barg and provided her with lots of help and advice, and are working with her to create the cocoa products she needs to make great raw chocolate, including Mott Green at Grenada Chocolate and Frederick Schilling at Big Tree Farms. Others in the greater fine flavor industry—notably those who manufacture at origin— have responded to the increasing demand for raw chocolate, relishing the challenge to get rid of the bitterness and acidity of raw beans and still gain flavor without the heat of traditional processes. Pacari in Ecuador has and continues to produce raw chocolate that is well received by a wide audience.

Pacari can also put some additional safety guarantees in place as it controls the entire process at origin.

But no other fine flavor manufacturer has ventured into raw with the focus, dedication, and eye to the future as Big Tree Farms in Bali. Before it opened its landmark bean-to-bar bamboo chocolate factory in Southeast Asia in 2011, Big Tree Farms had already turned to raw from a value-added perspective. "Obviously, the more you process a product, the more value there is to it," says Big Tree Farms partner Frederick Schilling. "We started by doing some powder and butter pressing and found the niche that would maximize our efforts: the raw food market. We became pioneers for a cold-process powder and butter. By doing it in a cold-process manner, we preserve a much higher nutrient content. And we can't produce enough to meet demand."

When its bamboo factory was up and running, the first bar Big Tree produced was a cold-processed, unroasted raw chocolate sweetened with the company's coconut sugar. "It is so good," says Frederick. "Look, I'm a chocolate snob. I admit it. Most of the chocolates on the market I cannot eat. In fact, I'm kind of bored with chocolate, to be honest with you, and that's why I had to do something new. But what is great about the stuff we're making is that it is not your quintessential chocolate flavor. There is a whole different flavor—a very warm flavor, a very soothing flavor . . . very different. I love it."

Vanessa Barg of Gnosis would agree, of course, and her goal for the future of raw chocolate is no different from Frederick's and indeed all the fine flavor manufacturers we spoke to: maintain her company's integrity, educate consumers about world of fine flavor, and build the market for fine flavor chocolate as a whole. Her love for raw chocolate may remain exceptional in the near future. Certainly more consumers and manufacturers will unite behind David Castellan's sentiment that "people have been using fire since man has been around so I don't know why we need to stop now." But Gnosis is expanding to meet demand and grow its product line and flavor profile. Hundreds of stores devote considerable shelf space to her chocolate from the obvious candidates (Whole Foods and health and nutrition stores) to the surprising (Hy-Vee grocery stores in the Midwestern United States). The bars sell for fine flavor prices too: between $7.99 and $8.99 each. Simply put, raw chocolate is evolving in many ways, and while it will always be a niche in fine flavor chocolate and the industry as a whole, the promise is there and it shows no signs of slowing down.

But while Vanessa and others are making a fine and flavorful future for raw, organic chocolate, neither she nor Frederick Schilling's cold process—not even David Castellan's proper fire—could raise the flavor bar for many certified organic beans.

Organically Speaking

Like fair trade, consumer consciousness is shifting toward organic certifications, which they see as more just and better for the environment and themselves. Every indicator shows growth in consumer demand for organic-certified products. Throughout the recent global economic downturn, customers worldwide have been willing to pay a premium for organic products, including chocolate and the milk and sugar often found in it. According to the Food and Agriculture Organization of the United Nations, demand is high for organic cocoa and chocolate.

Most fine flavor manufacturers have nothing against organic beans in principle. Organic certification is a less tangled web than fair trade: The International Federation of Organic Agriculture Movements (IFOAM) provides a market guarantee for integrity of all organic claims, and the Organic Guarantee System (OGS) unites the organic world through a common system of standards, verification, and market identity. Now that the United States and European Union organic standards are reciprocal, organic certification is well on its way to a more universal meaning on millions of packages worldwide. There are some problems on the ground—for example, organic certificates can sometimes be "bought" in corrupted countries. But overall there is less controversy surrounding organic certification. The certification is more transparent than fair trade and it is good for the environment.

Despite all this, organic cocoa commands a very small share of the total cocoa market—less than 0.5 percent of total production. The latest Organic Monitor research study commissioned by Fairtrade International estimates that the organic cacao market is only expected to grow 2 to 5 percent through 2013, with the major importers in North America and Europe, like United Cocoa Processor, Dagoba Chocolate (a division of Hershey's), Pronatec, and Barry Callebaut, accounting for most of the purchasing.

The problem is not demand then, but supply, especially the supply of organic fine flavor beans. This might sound strange given that more than 80 percent of certified- organic cocoa comes from Latin America, Madagascar, and Tanzania—regions known or developing a reputation for fine flavor. So why is supply, particularly fine flavor supply, not growing in line with consumer demand? The answers are—as usual when it comes to chocolate—a little messy, but for most people it comes down to two familiar words: farmers and flavor.

Like fair trade-certified beans, farmers are not paid much of a premium for organic beans—about 10 to 15 percent on average, though the highest quality can command up to 50 percent for their fine flavor profile, according to Curtis Vreeland, researcher for Packaged Facts. And then the question is, like premiums paid for fair trade beans, how much of that reaches the farmer? "The problem is this theory that the consumer pays a premium for organic and that somehow trickles down to the farmers and their workers," says George Soriano of Sibú Chocolates in Costa Rica. "But we know that works in theory, not in practice. It takes a lot of effort to produce organic on a larger scale. Farmers have to figure out what kind of fertilizers they can use and then source them and bring them to the farm. That's expensive."

Expensive not only in terms of certification costs but also time and immediate impact on those farmers' livelihoods. To get organic certification, farmers must stop growing their cacao using any inorganic pesticides and fertilizers for three years, complying with a complex set of standards (check out the USDA website to get a sense of the head-spinning conditions). Then they must complete a massive amount of paperwork. Only then can they pay for the privilege of certification, which can run more than $10,000 per year.

With small family farms growing more than 90 percent of the cacao in the world, the reality is that most farmers cannot afford the process of organic certification, and unlike fair trade there is less support from NGOs and other groups to help farmers get it. It should not be surprising then that some farmers who choose to grow cacao at all, choose not to certify. In the end, though, most small cacao farmers are doing little environmental damage because they can't afford the fertilizers and pesticides even if they want to use them. Jorge Redmond of Chocolates El Rey speaks for many in the industry when he says: "In essence all cacao in Venezuela is organic because the growers cannot afford the chemicals." Like many manufacturers, El Rey wants to create more organic products, but for now they simply try to educate consumers that most of their cacao is what some in the industry call "de facto organic."

But even if more farms do get organic certification and supply increases in the future, consumers should not expect a corresponding amount of organic fine flavor chocolate to hit the shelves. One might think organic chocolate would look and taste the same as or even better than conventional chocolate. But simply put, like fair trade, organic certification has nothing to do with flavor. Richard Callebaut calls this perceived connection to quality and flavor a "big misunderstanding" among consumers when it comes to certification. Mark Adriaenssens, Barry Callebaut's head of R&D in the Americas, builds on Richard's response to explain one reason most certifications don't work when it comes to flavor and cacao: The certifiers don't come from the world of cocoa and their standards don't reflect the complexity of how it is produced. "Rainforest Alliance comes from banana plantations. UTZ and fair trade come from coffee. And we struggle to translate this to cacao farmers," says Mark. "Small cocoa farmers are not comparable to big banana plantations. Look at organic cacao. We know that very few pesticides are used on these farms. So does organic add the value? Not if there is no quality behind it and nothing in certification is about a better flavor or better quality cacao. And organic is often just not good. We had to refuse half the lots when we first started with organic cocoa beans. It was not a sustainable situation. We had to train organic farmers how to do good fermentation and drying. It is better now. We know we can do better."

As a result, Barry Callebaut is developing its own programs in places like Tanzania for organic production, in which it controls the entire process. Of course, a multinational manufacturer like Barry Callebaut has the money and people to invest in and support those large-scale organic endeavors. No wonder most certified organic farms, fine flavor or otherwise, are the large farms and plantations that make up the smallest part of the cacao industry. Small farms that are certified organic are often part of cooperatives, and as previously discussed, many of those cooperatives often mix beans and/or ferment them poorly, producing a poor-quality product.

"We often can't get the right flavor from some of the specialty certified beans," says Gary Guittard, who wants to do more organic blends. "It is limiting in terms of the flavor that we can deliver with it because you don't have a lot of the types that are attainable and organic." On his website, Frank Homann of Xoco in Honduras calls organic "desirable from a policy standpoint" but immediately states, "Organic cocoa is not the same as fine or flavor cocoa." Xoco is supporting organic growing methods and starting a certification program for its growers, but states "there still isn't evidence between flavor and the growing inputs." Angelo Agostoni of ICAM in Italy sums it up most succinctly for those in the fine flavor world: "Yes, organic. Yes, fair trade. But first and foremost good chocolate."

All this is a big reason some manufacturers, even if they grow or use organic beans, find the whole organic movement sort of overblown.

Patrick Roger calls organic "a complete joke," and although he buys organic and fair trade products, he does not advertise this fact because there are just too many certifications to keep track of, and because of just how bad some of it tastes, which would actually turn off his customers. "Today, especially in Paris, people are willing to pay more to have organic . . . because very often, it is the Parisians who have the notion of the beautiful and the good. But not long ago, I bought a piece of organic chocolate, and it was disgusting," Patrick says. "That doesn't necessarily mean that *all* organic chocolate is gross. But the label in my shop for my customers means maybe two or three more people will buy the chocolates. It has little value to me."

"Organic used to mean that there was a farmer behind it who cared," says Mott Green of Grenada Chocolate, which is in fact a 100 percent certified-organic cooperative. "Now it means what the US government or the European Union says it is. Only the large industrial factory farms are the ones who can easily afford and complete the level of paperwork required to certify organic or anything else." Felix Inderbitzin, Felchlin's cacao sourcer, feels the same way: "Every big company is now searching for a certification of some kind. For a small company like Felchlin, or even smaller manufacturers, it's complicated to get all the certificates. There are hundreds of them. Working to get the right maintenance on the trees and the right plan in place in the fields and making sure everything is protected . . . that's more or less how we are behind the sustainability of cocoa and how we help the farmers."

Rick Mast of Mast Brothers Chocolate in Brooklyn, New York, thinks certifications have their place but only as an important part of industrial manufacturing. "If chocolate is being done on a huge scale, then to ensure things and attract customers, you need certifications," says Rick. "I look at soy lecithin the same way. It facilitates industrial manufacturing. But industrial-scale manufacturers can educate their communities. They can try. But their way is through a little stamp." That's why Mast Brothers, even though it uses organic chocolate, intentionally puts *nothing* on its hand-wrapped chocolate bars but name, origin, ingredients, and cacao content. "It's about showing instead of telling for the future. We don't like certifications. It is a bunch of mumbo-jumbo crap. This is a simple process," Rick continues, gesturing to their machines. "Look how simple our factory is. That feeling of simplicity has to be spread everywhere."

Truth is, Rick, Mott, and Felix are all right: No certification is going to do more for farmers than increasing yield and helping them get double the world market price for great beans, grown and processed well. If those beans are organic, the premiums could be quite high, but for now those fine flavor beans remain a distinct minority when it comes to organic.

But things are improving. Angelo Agostoni is pleased that the yield and quality of the organic beans ICAM uses have increased steadily. "In the early days," he says, "quality was not the most important thing, so the customer would buy because it was organic and supposedly good for the environment or growers. Quality has now matched the claims. But I am not one of the ideologues. I don't believe that if you live on organic food you will live forever or solve the world's problems. But organic farming has proven to give very good results. Our experience is that well-managed organic plantations—just because the organic certifications require the farmer to do more work on the plantations—may result in better yields than conventional plantations that are treated with pesticides or standard conventional treatment." Angelo cites The Green Cane sugar cane project as an example of how this works, which in its thirty-year history in organic farming has improved yields and financial return on its sugar cane.

This is certainly happening at certified organic cacao processors and manufacturers who work at origin. For Mott Green at Grenada Chocolate, his organic point of difference is all connected to the quality his egalitarian cooperative produces. "I like to call us the first 'organic gastronomic' manufacturer, or 'gastronomic organic,' if you prefer. We've come to a place in Grenada where we have the best cocoa in the world and certified it. If we have one special niche, it would be our ethical side and this very pure alternative to all the bogus fair trade-certified stuff. People can come right here and talk to a farmer and see for themselves. We have that kind of transparency."

And some manufacturers are all for the certification process. Hugo Hermelink, who owns FINMAC farm in Costa Rica, is no stranger to the work Angelo describes or Mott's transparent standards, and he has been at this a long time. Hugo switched his farm to organic in 1999 (certified in 2002) to get better prices. He later added Rainforest Alliance certification and the first Latin American UTZ certification for the same reason: he knows the certifications are important to his customers. And because Hugo produces a high-quality fine flavor bean that is processed on the farm, he stands out as a model for other plantations. He also sees no choice in keeping the certification train going: "It is expensive to certify a lot of work and a lot of paperwork and it is not fun to do. But we need to keep going. In 2010, we had a major insect problem like never before since we've been organic. We couldn't control them. We tried eight organic products to control these insects and nothing worked. Finally, the rain controlled it, but it can come back any day. We lost at least forty percent of our harvest to these insects." Sure, Hugo knows some certified organic farms could use pesticides and herbicides and still be organic if those farms demonstrate that they tried organic methods first, but he did and will not despite the risks like those insects. He prefers a natural, aggressive management approach on his plantation now and for the future.

Santiago Peralta, whose Pacari Chocolate in Ecuador has long been certified organic (as well as kosher), has gone beyond organic and certified as biodynamic in 2012 through Demeter International—a process that took five years to complete. Biodynamic agriculture takes sustainability and organic farming and manufacturing at origin to another level by considering the farm holistically from soil to plants to insects and animals and is widely regarded as the highest grade of organic farming in the world. Santiago also sees it benefitting farmers in the long run, even if buyers for their beans change, and hopes that this fact and his success will help other chocolate manufacturers see biodynamic agriculture as worth the extra time, effort, and cost:

"The nice thing about being biodynamic is we have seen productivity go up. It is very precise and the results command a price and deliver quality. This is already happening in wine. The most expensive wine in the world [Romanée-Conti in France] is biodynamic and many of the top-rated wines are biodynamic. Because it is not just environmental, it affects the flavor. After five years, this is really working. I can show it. I know it is not easy to become biodynamic and even harder to be certified, but I hope people keep'going in that direction, especially if it helps people understand why they should pay more for their chocolate and help the farmers."

But even as organic quality improves and, as Santiago hopes, biodynamic certifications grow in number, it does not mean manufacturers and chocolatiers will be able to or want to create the 100 percent organic chocolate bars and bonbons that consumers demand—even those who work at origin. That's often impossible and not even desirable when you consider the other inputs. George Soriano and Julio Fernandez of Sibú Chocolates in Costa Rica source chocolate from FINMAC but must deal with the fact that there is no certified organic milk in Costa Rica; they would need to import that. They prefer to support the local economy and use Costa Rican milk instead. Madécasse, which manufactures in Madagascar, is facing the same dilemma as it goes for organic certification at farm level; the company is uncertain if it will make the entire finished product organic since no certified organic sugar is available in Madagascar.

Manufacturers not at origin, like ICAM, can also be restricted in their development of new products because of the limited availability of quality certified- organic and fair trade products beyond the beans. For example, ICAM developed dark chocolate bars with dried sour cherries for a private label, what Angelo calls "a perfect marriage of flavors," and hoped to develop more products like it. But there were only two sources of organic certified cherries worldwide with the flavor profile the company needed. "In the end," Angelo says. "We were forced to buy an X-ray machine, as the quality control for foreign bodies was horrible." As a result, ICAM has switched its focus to developing tasting squares and Neapolitans as opposed to filled products.

All this messiness begs the question: How will consumers understand this, when consumers usually follow the herd when it comes to certifications? The answer is once again education—helping consumers understand that they are going to have to engage the companies that market products and know what their values are before they dismiss them for not having a stamp on their wrapper.

In 2003, David Castellan of Soma couldn't find brokers to sell him fair trade and organic beans. Now his inbox is filled with people pushing certified beans, and others constantly send samples. Ironically, after learning more about the fair trade and organic systems, he has less interest in those kinds of beans for the future and takes the time to address any consumer concerns, not through his website, but directly one-on-one. "Our solution is basically to put as much info on the label as possible. Let's say I want to make a Haitian and Madagascar blend. The Madagascar is going to be organic and the Haitian is not. That leaves me in a weird place with those consumers because I can't call it an organic bar but a lot of it is organic, including the sugar. That's going to be confusing to those consumers who want all organic, so you're going to have to let them decide. And education is the best way."

In fact, consumers may already be getting there. Anne Weyns of Artisan du Chocolat says her customers in England are starting to view all certifications with more "realism." As a result, she has actually seen demand for organic decline, not just because of the premium price it commands during a recession. She thinks the future lies in issues of sustainability and in more cocoa-growing countries processing beans and capturing a higher proportion of the value chain. This is even true in less-developed markets. George Soriano of Sibú reports an undercurrent among younger consumers in Costa Rica, especially those who work in the business world, who understand that part of the story and who have now become antiorganic and anticertification. "They don't want to believe it. They don't want to support it," says George. "They are very conscious of greenwashing in general and organic greenwashing specifically, and say they would prefer not to even worry about organic—that it doesn't add any value to them at all. What is important to them is the flavor."

In Australia, where Rainforest Alliance, fair trade, and organic products are all very popular, Michelle Morgan of Zokoko sees the same thing happening. She does not make an organic chocolate and thinks people are coming around to understanding why: organic chocolate doesn't taste that good despite what they've heard. "Consumers are starting to ask questions, and if I can answer them and they are satisfied that what we are doing is okay by their standards, we can go beyond certifications. That's why raising awareness is key. Because it comes down to flavor and what is certified is not necessarily good, I cannot support and I cannot be part of that system. I am not just going to go out and buy a bean because it has got that sticker on it. In the end, I need beans that have been cared for properly. I need to know that the growers are getting fair return. I have to sell a chocolate that I am happy with from go to whoa."

"Go to whoa" is the wonderful Australian expression for beginning to end—one that is particularly appropriate for where we are both in this part of our journey and the book as a whole. And while the consumer has dictated a lot of what has been covered in this part, the next part is where the consumers' pleasure remains paramount to the fine flavor industry even as their likes and requests do not. Adapting to trends and cultures is one thing, but fine flavor manufacturers and chocolatiers see themselves as the driving forces now and into the future. Pierre Hermé says it as straightforwardly as anyone: "I do not follow what my customers like or ask for. My own desires and inspiration shape my range of products."

Hermé's Parisian counterpart, Patrick Roger, would agree but clarifies the relationship further: "In my shop, the primary client is *me*. I am the one who must guarantee the quality of the work. I dictate the taste of the chocolate, not the clients. However, it is the clients who inspire me. It is a love story."

PART FOUR

Performing Flavor:
The Art of the Chocolatier

When you enchant people, your goal is not to make money
from them or to get them to do what you want
but to fill them with great delight.
—Guy Kawasaki

A Rocher, Junior Mints, and Some Space Dust

Michael Recchiuti of Recchiuti Confections in San Francisco is really into his Junior Mints. Anne Weyns of Artisan du Chocolat in London has customers clamoring for Space Dust popping candy. Patrick Roger in Paris proudly proffers rochers that share part of the name of and bear at least a passing resemblance to the popular Ferrero Rochers. When asked for representative examples of their cutting-edge craftsmanship that give a sense of where they are going, three world-class chocolatiers mentioned products that don't look like traditional bonbons but riff on confections well beyond the world of fine flavor.

But, of course, the results are decidedly fine.

Michael Recchiuti's PEPs, or Peppermint Thins, might be shaped a bit like Junior Mints and come in a green and white box, but his version features a fondant of organic Willamette Valley peppermint and a shell of his 64% custom-blended semisweet chocolate. Anne Weyns is not putting popping candies in her palets d'or but coats the popping candies in chocolate and then uses them as the base of her Space Dust UFOs—chocolate molded in the shape of a UFO that melts and explodes the moment you put it in your mouth. And the rochers Patrick Roger offers? They do come in milk chocolate, have a somewhat grainy praline that includes hazelnuts, and feature a rough exterior similar to the gold foil-wrapped Ferrero Rochers from Italy (*rocher* is French for rock, and not an unfamiliar term in chocolate making). But no one could mistake Patrick's cube-shaped rochers for the other in a look or taste test.

So why these products, and why place them at the start of this final part of our fine flavor journey? Are we sending a message about the future? "Watch out, Cadbury, and heads up, Hershey's, the chocolatiers are coming—upscale is going down market!" Hardly. But it cannot be too surprising that creations like fine flavor Junior Mints, popping UFOs, and ritzy rochers are top of mind for some chocolatiers when contemplating the future. The worldwide chocolate market may continue to grow in spite of global economic turbulence, but times are still tough, and, especially in major chocolate-consuming regions like Western Europe and the United States, every segment of the market is looking to create deeper connections to its customers— to become part of everyday life, not an occasional indulgence. As the *Wall Street Journal* reported in a November 2011 article "Breeding a Nation of Chocoholics," middle-of-the-road multinational bonbon brands like Godiva are already aggressively adapting, creating new packaging such as individually wrapped bite-size pieces, and pushing hard into supermarkets in order to make their chocolates everyday snacks.

For the fine flavor chocolate industry, this is good news; with broader exposure to more midmarket brands like Godiva, Lindt, and Ghirardelli, consumers will appreciate a greater range of chocolate and be educated up one level from candy.

And let's face it, a little playfulness and comforting familiarity go a long way to welcome new customers, remind current ones to have some fun, and delight everyone. Anne certainly sees her UFOs, gold-wrapped coins, and similar products in her store as a blow against the "snobbism" that pervades some of her fine flavor chocolatier world. Michael freely calls himself a "Junior Mint freak" and admits he created his

PEPs, as well as his upscale peanut butter cups and s'mores, because he loves the junk food and candy he grew up with. And certainly Patrick Roger is no stranger to using chocolate as vehicle for both playful performance art and the preservation of the past. Look at the giant endangered species sculptures created out of chocolate that have decorated his shop windows: penguins, polar bears, elephants, and gorillas, oh my.

In fall 2011, we found Patrick busy hand-carving hippopotami from a solid bar of chocolate seven meters long. Those hippos took almost a year to finish before claiming their place in the windows and becoming literal gateways to his bonbons—a lighthearted welcome with deeper resonance connecting past and future. "My customers always have the impression that when they come to my shops, they are tasting the taste of their childhood," Patrick says, noting this connection is deliberate and profound and hopefully extends to today's children as well. "Children eat like us. The mistake is to not give them chocolate to eat. I understand that it is a question of money, but that's not really the point. My daughters heard music when they were in their mother's womb to appreciate music at the earliest of ages. Now the taste buds form between zero and five years, right?"

France has a "culture of taste," Patrick argues—a much deeper cultural connection to flavor than, say, the United States or developing countries. But he knows the world has changed and he sees his chocolates as part of a future that reclaims and preserves that French tradition by understanding the importance of taking a homemade approach. "The goal is to rediscover the excellence of taste again, on a cultural level," he says. "Look at the cooking that is done today—the premade mixes . . . thirty years ago, premade mixes did not exist here. We mix everything from scratch here and thanks to some of my colleagues and me, we are getting back to this type of cultural taste."

Like Patrick, Anne Weyns is interested in fine chocolate products that connect not only to childhood but also more broadly to family, flavor, and future—a way to step up to more sophisticated tastes and reach back to happier times. "I don't think it is just the taste of the children but the taste of the parents," Anne says. "Most of our customers are adults, but they have kids, and we just did not have a range for children because we didn't think that parents would actually want that. But the more that we produced products like the UFOs, the more people were really into them. Now, whether people are actually buying them as nostalgic treats for themselves or they are actually giving them to their children I do not know. It could be a bit of both. Still, we just cannot produce enough. It is just something that is a little bit fun and different. People like to be surprised and reminded of when they were growing up."

But don't let this aura of childhood innocence fool you. While these products may be playful they can be just as difficult—even more difficult—to create as a traditional bonbon recipe. For example, from creating test batches to ensuring some kind of shelf life without the use of preservatives found in Junior Mints, Michael Recchiuti spent nine months perfecting his PEPs. That Junior Mint may be far removed from Patrick Roger's "culture of taste," but Michael sees some similarity in reclaiming the true origins of the candy he loves: "I think the initial take on the Junior Mint and peanut butter cup were pretty damned good. Somebody made it on a small scale and then the big companies just built machinery to make it happen and replaced everything with hydrogenated fats, but they were good ideas. Those are things I'm kind of into these days."

In fact, Michael could be speaking for many chocolatiers when he talks about the philosophy behind the creation of those PEPs. "Anytime something becomes trendy, I go back," he continues. "When I got started I was really sort of into the herb and tea infusion and different flavors. And then it got to the point where everybody was just trying to be weirder than the other. And so I just kind of shifted because I thought I don't really want to be a part of that, and I started making things I like."

Chasing good ideas, making what they like, and delighting their customers: three things that united every chocolatier we spoke to about his or her current work. Chocolatiers may be in very different parts of the world, sell to different audiences, enjoy different ingredients, and use different origins and custom blends of chocolate, but whether talking about the creative process, customers, packaging, or building a business, they are unified in their passion for delivering quality and flavor, and in the pursuit of ideas on their own terms.

Not because these chocolatiers want to be stars, though. They just want their chocolates and bonbons (and the experience of eating them) to be the best they can be. We may not know their faces, but their names—usually on the door—and presentations command a premium. They guarantee quality, luxury, and a singular human vision behind every bite. First and foremost, all are deeply connected to education about and the preservation of taste. In this way, they embody the spirit described by Michael Ruhlman in *The Making of a Chef*: "The chef hadn't used the potato as a basis for displaying flashy, flamboyant skills but had placed his skills in the service of the potato." Fine flavor chocolatiers are in the service of the flavor of chocolate and never lose touch with that flavor and all it is capable of conveying.

Of course, most chocolatiers know they need a little flash, flamboyance, and fun to stay motivated and survive. They also know they must balance their artistic and flavorful pursuits with the continued production of the bonbons and other treats their customers expect to find and have grown to love. Whether watching over those creations, traveling the world to discover new pairings, or simply taking their love of Junior Mints to the highest level, fine flavor chocolatiers are all deeply aware of the "stage" they work on and the importance of taste in every performance.

Getting Fresh and Playing with Food

Twenty-five years ago, Bart Van Cauwenberghe had just finished a stint as an officer in the Belgian army and was selling vacuum cleaners door to door in his hometown of Deinze. If he imagined taking the stage, it was as a drummer extolling the virtues of rock and roll in a local club. That was then. Now, Bart often takes the stage hundreds or even thousands of miles from Deinze to extol the possibilities of freshness and flavor in chocolate as he did for a packed house of culinary professionals at the Scottish Chefs Conference in November 2011.

How far he and fine flavor chocolate have come in a generation.

Bart relishes the opportunities to do cooking shows around the world as part of his work as a Belgian Chocolate Ambassador for Barry Callebaut. And he is careful to use the word "cooking" to describe what he does. That's because Bart knows it is essential for his craft and its future that chocolate be understood and explored as a fresh food connected to and capable of being paired with almost any other food, be it an herb or spice, fruit or vegetable, tea or coffee, heirloom tomato, goat cheese, foie gras... Bart has explored all of these combinations and more at his two-decades old shop, De Zwarte Vos, in Deinze, about an hour north of Brussels. But he says his real search for taste and all its future possibilities started when he prepared to teach his first cooking classes five years ago.

"I read every cookbook searching for ingredients. I spoke to professors," Bart says. "I spoke to guys who make spices and herbs and asked for coriander, and when the guy said, 'What kind of coriander do you want?' I had no idea what he meant. He said he had more than thirty different corianders. He had 700 different types of pepper. I devoted all my spare time to researching these flavors. Today, all my classes are built around these flavors and how to create nice, balanced chocolates and other dishes with chocolate. That is what we must do for the future. Today, we just eat. We just do, instead of thinking about what we are doing, creating, and eating."

Bart then explains how he puts together "tasting boxes" of bonbons and other ingredients for classes like the one in Scotland, urging the audience to rethink what they know and explore different possibilities. "What if I say to you, 'Béarnaise sauce—what is that sauce?' you answer, 'it's creamy with some spice and tarragon.' Now why not put that in a bonbon where first comes the cream, then the spice, and then the tarragon finishes. That's how a recipe delights and surprises. But you must be giving them an experience, so I have to think about texture, too. If I give you a chocolate mousse as big as a beer glass, you will take two or three bites and be full. But you're not full; you're bored. So what if a few spoonfuls down I put another layer where it's crunchy? And what if there's another layer after that? That's exciting. You want to keep tasting. Not eat, *taste*."

If Bart's passion for taste and craftsmanship sound decidedly chef-like, his confession of where he wants to go next will come as no surprise: He wants to take chocolate beyond the bonbon, cakes, and pastry and open a restaurant—all with chocolate. "I'll make fish with chocolate. I'll make cheese with chocolate. I want to take chocolate out of the dessert corner," Bart says. "When people taste my fish with chocolate they don't say 'What is this?' They say, 'That is really nice. That is delicious.' Maybe I give them a dish with smoked trout. What? Why not? Why not connect the flavors? Chocolate can go there. It is a vehicle for more flavors. Chocolate can go with fish, so then put spaghetti in the dessert. Have you ever boiled spaghetti in apple juice? Why not? The spaghetti has no flavor of its own. Spaghetti with the flavor of apples and then cinnamon ice cream on top of it . . . believe me, everybody will understand. Everybody will like it."

Patrick Roger does. He used a similar pasta analogy when he talked about the form his chocolate takes as the vehicle for taste: "It is like pasta. You have shells, spaghetti, macaroni—it's the same recipe, but not the same taste."

Michael Recchiuti explains this expansive versatility further by comparing it to a passion he shares with Bart: drumming. "You sometimes need restraint as a drummer," Michael says. "You have to hit very lightly in orchestral music, but when you're playing rock and roll you can kind of just get there and jump in it. I apply the orchestral philosophy to making chocolate. I'm using really good chocolate so I want people to taste it. Maybe if I bought really crappy chocolate and bad beans I would mask it with alcohol and all kinds of crazy flavors. But I'm using good chocolate, so let's just find great ways to taste it. That's what Patrick Roger does. His whole style is high-acid great chocolate. Guys like that know how to pay attention to the sensitivity of your chocolates."

Michael will explore this idea on a larger scale when he opens his Chocolate Lab café along with a retail outpost called The Little Nib in his original Dogpatch neighborhood of San Francisco. Some people feel he is going off brand by opening up a café, but Michael sees it as a logical extension of who he is and where he is going for the future. His goal for the café? Do some really fun things and avoid being trendy by doing the things he likes; let the performance come from the flavor. "We won't do really fluffy, crazy elaborate plates of desserts with 750 components," Michael says. "It is not going to feature decorative chocolate pieces that are going to fly off the plate when you knock it over and hit the person next to you and hurt them. I can't get into that anymore. I think visually it is fun but is not what I'm about and what this future is about. We'll say, 'Here's a really good slice of this.' And you're going to eat it and love it and cry. That's what I want. I'm just dialing into flavor."

And Michael is helping his customers dial in, too. The possibilities for education on a broader scale are not lost on him. "We can actually make things for people like vertical tastings and walk them through it," he adds. "We can taste a particular Madagascar chocolate in three different formats but keep all of them true to the chocolate so there's no disruption. There might even be a sorbet or an ice cream. These kinds of tastings are very effective. People really get it."

Thomas Haas of Thomas Haas Chocolates in Vancouver certainly understands this drive for new and deeper experiences that put flavor before flash. "What I always look for in food is a good experience," he says. "What I don't look for is excitement that disappoints me." And what disappoints Thomas, who gained a rich appreciation for cooking as Daniel Boulud's pastry chef in New York City and has made chocolates for some of the finest hotels in the world? "I can rarely give my customers the best of the best because the moment it is made, it starts aging," he answers.

Given that almost all small and craft chocolatiers make their bonbons fresh with cream and other perishable ingredients, their bonbons have shelf lives of about two or three weeks and should be stored in a cool place (but always eaten at room temperature). No wonder Thomas Haas thinks it is harder for most chocolatiers to provide freshness and turnaround than to come up with new recipes. "If you have good ingredients and a recipe that works, and if you're not out to lunch when you mix it, then you shouldn't screw it up," Thomas says. "I know how amazing it is when we make our chocolates today and I can just put them out in the store where there are boxes lined up to be filled, and I know that people are going to buy them and have them that night and it can't be any better. For me being able to provide that more and more and more consistently would be the goal I would like to achieve over and over."

Pursuing the ability to preserve this freshness in order to build a bigger customer base in the mass market has Anne

Weyns of Artisan du Chocolat chanting "chill, baby, chill" to her local supermarkets. That's because supermarket candy aisles—where almost all chocolate is found—are hardly cool and their bins are barely covered. Without the preservatives and stabilizers added to most shelf-stable box chocolates, fresh chocolates would go there to die, and there aren't any alternatives—a fact Anne finds curious: "Most other categories have an alternative premium-equivalent in chilled. Take soups. You still have canned soup but you also have hot soup, frozen soup, and fresh chilled soup in the refrigerated area. Those are sold at a premium. The only two categories that have never done that are boxed chocolates and baby food. Which is strange because you would think those are two things people really care about."

But Anne thinks that as customers demand fresh alternatives to shelf-stable supermarket chocolates, one of the more interesting future developments will be the movement of boxed chocolates and bonbons in the mass market from their usual shelf space in the candy aisle to the chilled space of desserts. She believes this is now possible because customers in London—and indeed worldwide—continue to be educated about fine flavor chocolate and understand more about chocolate as a food. As a result, they are getting used to the idea that the shelf life should be shorter. "Ten years ago in the UK there were not many chocolate shops, so people could not really understand," says Anne, who along with Paul Young, William Curley, Rococo, and others, have ushered in a huge chocolatier quality wave in London. "Now they understand that if they go to a small shop, they get chocolates that are fresh. If they go to a supermarket, they don't."

Anne, who has two Artisan du Chocolat shops in London and sells through the three Selfridges department stores, has already discussed this change with retailers, like Waitrose, in the United Kingdom but there are hurdles both logistical (the stores would need to replace something that is currently refrigerated and selling, add in new refrigerators, or move boxed chocolates next to desserts) and territorial (the chocolate buyer doesn't buy chilled products and the chilled-products buyer doesn't have shelf space—and just try to get them to work together). Fresh chocolates almost need to be a new category unto themselves.

Up until now, progress has been slow, especially as the bigger brands with the most marketing money in Europe like Lindt and Godiva are content with their shelf space and placement. Only a few stores in Belgium and France have tried the fresh alternative. "Retailers also worry that people won't understand what these chocolates are," Anne adds. "With chilled space being at a premium people generally don't take a lot of risk . . . certainly not until there is a more positive economy, as retailers in recession tend to stick to categories already established."

Nonetheless, Anne believes change is inevitable. "We are still confident it is a natural evolution. It is just a matter of the right time, right people, right circumstances. If supermarkets can have sushi, there is no reason they cannot have boxed chocolates that could last up to a month. I think Waitrose is about to launch a range of truffles made with cream and fresh ingredients that has a shelf life of six weeks, which is much, much less than the usual six months to a year. And we can get around this by offering them seasonal products in mid-November that by end of

December will be gone—that have shorter shelf lives of maybe four to six weeks, so not the freshest but still fresh enough."

"Fresh enough" may work to expand the bonbon market in established flavor-consuming regions like Western Europe, but freshness overall is working for chocolatiers in establishing new markets for the future. In China, "fresh enough" chocolate would be more than most people have ever had access to, let alone appreciated. This is what Laurier Dubeau and his partner Polly Lo faced when they launched La Place Collection in Beijing in 2005. Beijing may be an international city, but it is still what Laurier calls "a big village where people sit in the corner and eat noodles and wonder how come our chocolate is so expensive because I can buy a bowl of noodles for the cost of one piece." In fact, when La Place opened, its customers couldn't find any fresh dairy products, so why would they desire something with fresh cream?

But Laurier and Polly still felt freshness would be key in winning over customers. And things did change quickly as customers became very, very curious. Today, Laurier says, "I will tell them I'm doing a chocolate with thyme, banana, and walnut and they act surprised but then try it. And they might like it or not but they will try it."

What about shelf-stable multinational brands like Nestlé that are in China, or Godiva, which has an aggressive Chinese expansion plan? "We don't see them as competitors," says Laurier. "They help create awareness of chocolate. People will buy their chocolate and they will start to enjoy more chocolate so it's good for everybody. After that they start to compare just like wine. They try the Belgian chocolate companies that are here like Leonidas and Valentino. These are not fine flavor companies, but here they look like premium chocolate. If they like it, they come to us and can compare, and they like ours better and say, 'Wow, this is really fresh,' and ask questions. That's when we can explain to them that it is fresher and that we don't import it and that the shelf life is shorter and we take care of the chocolate. . . . They are starting to learn more of the difference."

Freshness is even top of mind for chocolatiers in flavor-growing but less-established flavor-consuming regions like Costa Rica, where George Soriano and Julio Fernandez of Sibú Chocolates see it as a valuable point of difference in educating and establishing their market. "Things don't move as quickly in Costa Rica," George says. "We get awards here for being innovative, but Julio and I know that we are just innovative *here*. People are not really used to eating fresh chocolates and fresh bonbons here. They are used to eating Lindt or a Whitman's Sampler and throwing away half of them because they don't like them. Ours are all interesting and different to them."

Of course, that same freshness also makes Sibú's bonbons, like many other fine flavor bonbons, unexportable. But George and Julio have turned this into an advantage. By leveraging their local position and tapping into regional tastes by using fresh ingredients that are largely indigenous, familiar, and appealing to Costa Ricans, they are producing chocolates in a way mass-market brands, and even many fine flavor chocolatiers, never could. "One of the advantages of our chocolate and bonbon business is we are here on the ground. That opens us up to exceptional fresh chocolate and tropical ingredients and a lot of great flavors that we can incorporate into our bonbons," George says, pointing out that they can work directly with Hugo Hermelink's FINMAC farm in Costa Rica to get different kinds of chocolate quickly. "We can create five or six different lines of chocolate because we are right here. We can make it in small batches at little extra cost. We can decide we are going to make 500 kilograms just with two trees and see how that turns out, and we could also do the same thing with fresh vanilla and fruit and spices and herbs."

Many chocolatiers can only dream of having Sibú's access to the freshest ingredients and chocolate, but others in cacao-growing regions are living the dream: A trend is slowly developing with small chocolatiers popping up in flavor-growing regions like Venezuela. But one thing still remains true from the jungles of Costa Rica to the streets of Beijing to the boulevards of Paris: regardless of their access and approach to ingredients, the chocolates being made are always reflective of the tastes and stories of the chocolatiers behind them.

Recipes for Success:
Pursuing New Directions and Ingredients

Despite the unconventional examples that opened this chapter, most chocolatiers continue to choose more traditional-looking creations to satisfy their personal passions, to pursue a great idea, and to explore the possibilities of taste. And while a shared vision to enchant and educate their customers unites many chocolatiers, the ingredients and textures of their creations are all over those spectra.

And rightly so. Fine flavor chocolatiers worldwide have different types of customers, different palates, and are at different stages in their careers as artists and chefs. There may be similar tastes in every shop, but any idea of there being one future for particular ingredients, flavor combinations, textures, or shapes is rubbish. What is certain is that each chocolatier has a zeal for chocolate—many getting exclusives and custom blends from manufacturers, choosing beans themselves, and touting the beans' origins on their websites, a trend that is only growing. And they have a shared fanaticism for finding and understanding the ingredients that go into their bonbons.

For some more established chocolatiers, the terroir of their homeland offers only so much, especially when the chocolate comes from places far away. For example, Patrick Roger is happy to snip herbs from his garden but says, "You need to explore and go further afield to find different flavors and not stick with just one for everything you do." That's why his pursuit of the best takes him to Corsica for oranges, to Ethiopia for coffee, and to Delhi for lemon. The latter is for a bonbon aptly named Delhi, which features an essence of basil and lemon in its almost caramel-like ganache and was the other bonbon Patrick pointed to when asked to talk about the future.

In Turin, Italy, Guido Gobino also believes the "best ingredients are found in places that are far from home," and he loves to find them and then combine them with his chocolate recipes for flavor and originality. His creative process makes him sound more like a marketing scientist than an artist or chef, but that is what it takes to achieve the best result: "After putting the new prescription to the tasting panel and various laboratory tests, I check the product's characteristics and decide whether to proceed with production."

What Guido describes is as true in Costa Rica and China, where there are few competitors, as it is in Paris and Brussels, where there are thousands. Despite varied business models and experience, fine flavor chocolatiers have the same attitudes toward creativity but work within clearly defined limits of personal preference and the bonbon itself. They take deep pride in what they make, and they want to share the story and process behind it, which is why almost every chocolatier wants you to taste while you talk to them. There's no better, or more delicious, way to understand where they are coming from and where they are going.

For example, George Soriano and Julio Fernandez of Sibú Chocolate in Costa Rica chose a dark milk chocolate bar flavored with coffee and cardamom and a dark chocolate-coated caramel infused with fresh ginger and coconut as their examples of where they are now and where they are going. The former reflects George and Julio's heritage (both are half Middle Eastern and like cardamom in their coffee) and the latter their country's culture of taste. "We call it the Costa Rican because it puts together all of the flavors that make up Costa Rican culture," says George. "Caramel based on sugar brought by the Spanish, ginger introduced by Chinese immigrants who came to build the railroad, and coconut milk from Afro-Caribbean cuisine, as many of the people on the coast are immigrants from Jamaica. Then it's covered in a dark chocolate shell with the cacao that comes from the indigenous pre-Colombian tradition in Costa Rica. The way that the flavors play together creates something new."

Of course, as George and Julio noted before, things move slower in Costa Rica, so a caramel, no matter how fresh, is hardly going to be the bonbon of the future most chocolatiers in the rest of the fine flavor world point to . . . or is it? For Anne Weyns at Artisan du Chocolat in chocolate-mad London, it is *all* about the salted caramel—a product she introduced in 2002. She calls the caramel "deceptively simple" with its chocolate shell filled with a liquid caramel that has added butter and sea salt. She also appreciates its adaptability. "We've done lots of extensions of it, like a salted caramel with fig and spices for Christmas," says Anne. "As a single product it reflects my sensibility and has been by far the most successful product we've ever done."

Meanwhile, half a world away in China, where customers have little bonbon exposure and would seem to be open to any possibility, salted caramels would be unique. Yet Laurier Dubeau of La Place Collection in Beijing does not and will not do one for the exact same reason Anne and George and Julio do: because he does only what he loves and believes is within his customers' culture of taste. "I don't do a lot of caramels and fondant because personally I don't like it. I do not add any more sugar to our chocolate. If it is too sweet, they won't eat it. Even if I do milk chocolate sometimes I find it is too sweet, so I add dark chocolate to make it less so."

Laurier struggles to pinpoint his bonbon of the future, because everything moves so fast in China. Given its size, Beijing alone could end up a slightly smaller version of fine-chocolate-mad Japan, where Pierre Hermé remains a legend and Hironobu Tsujiguchi's Le Chocolat de H packs customers in. According to Laurier: "Tastes are changing quickly and the Chinese are very curious about the world. The local customers who buy our chocolates appreciate them for the freshness and nonsugary texture and nonartificial taste. They love everything spicy. But everything is new to them. The raspberry peppercorn is something new to them. We are trying to use different kinds of teas and herbs. We just keep it natural and simple and go from there."

Given the speed at which things are moving in China and locals' willingness to try new flavors, Laurier thinks he could soon hear clamoring for bacon and wasabi chocolates from young people who want to try something new and adventurous. Whether he will indulge these flights of flavor fancy depends on his taste. Likewise, Thomas Haas of Thomas Haas Chocolates dismisses innovation for its own sake or shock value: "Do I want the future to be aerated ganaches or caramel and sea salt, which everybody does now? No. Do I think I am the person that people will look up in fifty years as the guy who changed the world? No. But if they say this is the guy who always made good chocolate? That would make me more proud.

"We just do what we love and what we think our customers will love. I'm kind of a traditionalist and classic in many ways in how we approach things. We spark it up on our dark chocolate ganache with food pairings, but we are very aware that so many—too many—chocolatiers are candy makers where it doesn't look, smell, or taste like chocolate anymore. If you want to be different for the sake of being different then, well, I just have no patience or taste for wasabi and mustard seeds and stuff like that. I'm sorry, it is not me."

Thomas does relish a flavor challenge but only if it appeals to him, and he gets his inspiration for future directions from whatever is around him. "Sometimes I think, 'It is spring so what do we think of?' Sometimes I'll ask around and try to inspire the staff so that they can think about it. Sometimes we think about flavors from the past or our childhood. Sometimes I will ask a customer," Thomas says, acknowledging that querying customers has its perils but still can lead to a new challenge or idea. "A customer will say, 'You know it is fall and I'm thinking of pecan pie.' And I will roll my eyes because that is what everybody out there does. You hear crème brûlée and lemon pie. . . . So I am thinking okay, how can I make something with that stuff and make it taste like good chocolate? What I'm thinking is vanilla and then pecan pies are kind of chocolatey but gooey so we can create a caramel with chocolate in it, add vanilla and have it bitter but also sweet with a little honey and then add small bits of caramelized pecans. Then, we try it, and if any of the people tasting it think it tastes like pecan pie and like it, then we are on the right track. That is part of the process in creating for the future."

In Pisa, Italy, Paul De Bondt and Cecilia Iacobelli of De Bondt Chocolate might agree. While they say their best product is "the one they have not invented yet," one of their favorite bonbons is one that a customer asked for, and the public cannot get. De Bondt has embraced the trend of creating exclusive custom chocolates for local clients using local ingredients. They recently created two for the Donnafugata Winery in Sicily, using its signature dessert wine Ben Ryé—one with ground almonds and figs that have been soaked in the wine and covered in a bittersweet chocolate shell, and the other made from a gelatin of Ben Ryé with a white chocolate ganache, also in a bittersweet chocolate shell.

Roger von Rotz, owner of the von Rotz Patisseries in Switzerland, is not surprised at this closer connection between customers and chocolatiers. He thinks the future of flavor is founded on the trust between the two, especially if the customer is another artisan. "As a chocolatier, I see myself as a guardian of cocoa culture, who offers the chocolate lover the chance to find the real and noble chocolates," Roger says. "I have found that I can work with all kinds of colors, fruits, spices, wine, liquors, and original couvertures—some more noticeable, others more traditional—and in these areas there is a huge variety of aromas and tastes, which are outstanding for premium chocolate. Depending on individual flavors, those customers can take a sensory journey with me through the world of flavors, from which they will come back with excitement. So consciously or unconsciously, everything in my head is about finding new tastes and creations that my customers don't already know."

But Roger cautions that he too will not experiment with what he calls "impossible variations." By this he means trendy ingredients that have everything to do with sales and nothing to do with his "joy and experience" and that spoil the taste of the chocolate. Recently, Roger was working on a new creation based on wild honey and another on tobacco, both of which inspire him and will help make his "flavor frame a little wider." He knows there will always be "chocolate lovers who are passionate about special pralines and lovers who are passionate about traditional chocolate. Chocolate will always be a taste thing! Our challenge in the future will be to show people the way that honors the culture of fine flavor cocoa."

This is exactly why even chocolatiers with the most playful or rebellious outward personalities like Michael Recchiuti, Patrick Roger, and Bart Van Cauwenberghe caution restraint when pursuing certain flights of fancy and flavor profiles.

"You have to think of it as a business you are trying to sustain," Michael says. "How many times are customers going to eat that blow-your-head-off bar? How long is it going to sit on a counter or in a cupboard after the first bites? You want people to buy your chocolate a lot. It is fascinating how many of the bean-to-bar things are happening in addition to the trend of people making rogue chocolates and those who are really into the kind of odd flavors like chilies and bacon. And that's fun and kind of edgy but for me it is: what are people really going to come back to and enjoy on a consistent basis?"

In this way, Michael, like Roger von Rotz and many others, feels that people aren't being true to chocolate. "I taste the chocolate and they are so flavor-heavy that you are not getting to experience the chocolate. They're interested in the different colors and shiny chocolate but they are not really interested in the taste," Michael says. He is especially disappointed when this happens with flavor combinations he likes, such as lavender, which he often finds too overwhelming: "We steep ours for less time because it has to be a whisper of lavender, not like we are hitting you over the head with a lavender shovel. That comes from restraint."

Michael hopes that the launch of his café will allow him to be a little less restrained in some of his pursuits. He has been playing with highly perishable products like water truffles that last only a day or two. "The reality is water allows the full flavor of the chocolate to come through. They have a chewier texture. It is almost like there is gum in them," he says. "The minute you add any kind of fat you start suppressing the flavor, but if you do it with water it is immediately perishable, so we add fat and suppress the flavor for texture and shelf life and unctuousness. Another thing we'd like to do are these agar agar [a Japanese seaweed jelly] stabilized chocolates. They are just water and agar agar in these little jellies and you'd taste it in three forms: its water form, and then its chocolate form, then maybe a liquid form."

The balance between innovation and restraint Michael seeks on that agar agar plate is exactly what Bart Van Cauwenberghe of De Zwarte Vos in Belgium thinks is most important in any future creations. "To shock people is really easy," he says. "To do chocolate with chile or pepper? People try it and say, 'Oh yeah, that's fantastic.' But if you are thinking about money long-term it is not good. Sometimes it's nice to shock people a little bit but if you can create nice balance in chocolate and whatever you do in life, in love, in food, whatever, then you are a happy man. That's the yin and the yang."

Bart honors this yin and yang in his own work by offering his customers a classic line (creams and nuts with familiar textures) and a trend line (flowers and spices or herbs and unusual textures). For his trend line, Bart has worked with foie gras, the tonka bean from Brazil, jasmine flowers, roses and raspberries, honey and rosemary, cardamom and orange . . . he works with flavors that compete with each other and flavors that complement each other. He understands Michael Recchiuti's desire to work with water, not just for the flavor of the chocolate but because figuring that out will help him figure out how to put olives or tomatoes in his chocolate. Bart also plays with texture to heighten the whole experience. For his Quattro, which riffs on the classic flavor of amaretto, the bonbon has four layers, all with amaretto, but each with a different texture to change the sensory experience. Bart designed a bonbon with chamomile and mint so the mint oil evaporates on your tongue and "you feel it when you breathe."

"That's how you play with flavors and textures so you can smell and even hear it as well as taste it. You have to think about all this, and that makes it so exciting," Bart says. "But I always say to people you have to taste first. Because if you just bite twice and then swallow you will never get it." This is one of the main reasons Bart does not label the bonbons in his signature Degustation Box: "People say, 'It is so nice but it is such a shame that we don't know what is inside.' But if I tell you and you know what's inside—that you have chocolate with vanilla or chocolate with hazelnut and chocolate with rosemary—what stays in the box? The rosemary. The one you never tasted before.

"I have to change your taste buds. That's why I put nothing on the top of the Degustation Box. I want people to degustate, not consummate."

In the end, all the chocolatiers we spoke with know that playing with this knowledge is not dangerous like fire but does carry responsibilities for the future. All have a deep respect for tradition and little patience for trendy flavor combinations unless they honor that tradition. They also share a disdain for fussy presentations, preferring layered and deceptively simple tastes. This preference made Patrick Roger question himself in the summer of 2011. "I went to visit a colleague of mine and I found that I could no longer eat my chocolates," Patrick says. "I preferred his chocolates that were simpler. I felt like Picasso. Picasso was a troubled man, right? I can be like that, too. What you find in art is very similar to what you find in taste.

"People love herbal infusions, but people also love those glass houses that are going up in places all over Paris. I live in a three-hundred-year-old house, made out of mud and wood. I know the glass house is more advanced, but still. But if there is something to see or taste, I will find it. There are very few people who have this attitude and who want to live like I do. And it is the same way with taste." That's why Patrick says he doesn't "search" for new recipes like those glass houses: "I might plan for, we'll say, two minutes, but I already know exactly what the taste will be—it's like kissing someone. It's another love story. Today, each person has his or her own vision of love. And taste is exactly the same way. That is how I am going to construct it. That is exactly what I am looking for in taste."

But almost as soon as a chocolatier like Patrick constructs a vision, chances are someone else is right behind him, or worse, watching closely. As competitors copy looks and flavors fast, leading fine flavor chocolatiers repeatedly need to innovate with new products and to develop and maintain their signature looks. How far can they actually go without compromise?

Familiarity and the Future:
On Differentiation and Connecting with Customers

Tiffany's blue box: Perhaps no package in the world is so distinctive, iconic, and longed for. That's because Tiffany's box makes a unique promise: No matter what is inside—no matter how big or what it costs—it came from Tiffany and no place else. Before you even get to the product and ingredients inside—gold, silver, platinum, diamonds, crystal—the box has delivered a feeling of quality and taste, thoughtfulness and love. Quite simply, the box separates Tiffany from every other jeweler in the world, making its value priceless. It is the pinnacle of distinctive packaging in the world of luxury goods. So much so that its robin's-egg blue color, a hallmark of the Tiffany brand since it was founded, is trademark protected in the United States and several other countries.

Is it any wonder then that most fine flavor chocolatiers think about their boxes and presentations in the same way? They want them to convey the same promise as Tiffany's: I will deliver on what you expect. In fact, for walk-in customers, especially in Europe, the store itself is like a giant box, conveying the vision of the brand and selling the bonbons and bars on display. All of this attention to detail is essential to maintaining distinctiveness in a world where imitation runs rampant and where even at close distance, the products look remarkably the same to most customers. European chocolatiers like Patrick Roger often carefully combine distinctive consistency and an element of surprise: Patrick has been using the same box since he opened and each store looks like a Patrick Roger and features the same products, but is different in its layout and has its own distinctive chocolate sculpture in the window.

But while distinctiveness in packaging is important to chocolatiers, innovation in that packaging is not top of mind for the future. Neither is environmentalism, largely because it has little value to the customer. In fact, Thomas Haas of Thomas Haas Chocolates was one of the few who mentioned packaging as he was working on a method of layering so that he does not need boxes inside a box to protect and present his chocolate. He also had some strong words for himself and his fellow chocolatiers: "We are all in the business hypocrites and I am one of them," he says. "We may be the only business out of 150 out here that has a full-on recycling program in everything that we do and we compost. But we still are always offering bags and giftwrapping and beautifully wrapped chocolate bars. This is hypocritical. I don't know how to deal with it. It is a part of our creativity, of course, but it is a shortcoming."

That said, packaging innovation is not top of mind with chocolate manufacturers either. Fine flavor companies spend just as much time considering the look and production of the wrappers for their bars as chocolatiers do with their boxes. And there has been lots of innovation on the environmental (recycled and reused as well as indigenous materials) and economic (both in terms of efficiency and local production) side in recent years. Yet most still use the familiar paper or cardboard wrapper with foil or plastic wrapping inside.

Some of this is based on cost and some on the fact that the chocolate itself demands so much of the manufacturers' time that there is no room to innovate. But for a food that is to be tasted and savored, only Chloé Doutre-Roussel presented something truly different: zip bags. "I think it is one of the most respectful ways to sell the chocolate because it protects it not only before— much more than the foil paper—but after," says Chloé. "It could last for up to two months once it was opened, though no chocolate should last for two months. When used properly, it really works perfectly. No chocolatiers use this maybe because it is not a very sexy packaging. I tried to make it less ugly, but it is highly functional."

But as discussed previously, when it comes to fine flavor chocolate, consumers are much more into form over function, how things look, and are often swayed by marketing. And when it comes to chocolate, and sales in general, there are always people who choose marketing and presentation over content and substance. Like the manufacturers, many chocolatiers see that as a problem for the future if fine flavor chocolatiers do not keep up and cannot match the marketing dollars. As Michael Recchiuti notes, "I've seen a lot of product where the boxes were good but the product was bad—and the reverse. That's always been a great disappointment for me just as far as people producing things and either it looks gorgeous and they don't follow through or is the most amazing product and the packaging doesn't reflect that. It's a disconnect."

Michael does understand, however, that packaging can be essential to representing the brand, the story of the chocolates inside, and the personality of the person who made them. Once his first line of chocolates was set years ago, he spent more time thinking about the box, the paper stock, design, and embossing than anything else. The result was lasting and profoundly personal. "A lot of our chocolates are about my story and my history in the pastry and food industry," he says. "The designer got ahold of all my old recipe books from the late 1970s and 1980s with all these coffee stains on them and scribbles and lines and drawings and letters to friends and thoughts and processes. He then scanned all of them and he came up with the idea to use them as a background on our boxes. He chose things that meant something to me and also looked great. We created a font based on it. It looks like how I think."

Roger von Rotz of von Rotz Patisseries in Switzerland also knows the game is changing with the appearance of new chocolates and chocolatiers: "This is not comparable to the past. Today, every chocolatier has to establish and maintain their own brand. We have been lucky but we know we must always be a work in progress. We are developing and changing very fast. This is sometimes hard. We redesigned in 2009 and now we are evolving in different ways. The aromas and tastes in the fruit and spice selections will develop just as spirits and wine did. We will realize mini chocolate collections with our exclusive chocolates. The ganache will get even softer and the chocolates will get flatter."

But Roger knows there is only so much he can do with taste and packaging and still maintain a distinctive yet traditional business: "Chocolate is still very traditional so there is only so much you can do before you go off brand. I think the cover/design of the place and the packaging helps with the perception of the quality of the product. We must make our products desirable to our customers, but if you design a product with earthy colors like we do, then it sometimes needs a little sales discussion to help sell the product. So there is no such thing as a bad future; you just have to adjust!"

From Beijing, Laurier Dubeau and Polly Lo of La Place Collection would concur, but for them packaging is building a future that they can adjust from. They also needed to make different decisions from those of chocolatiers in more established flavor-consuming cities. For example, at first their Chinese customers only believed chocolate that is made in Belgium or France was good. How then to differentiate La Place and still be what the customer wanted? As Laurier explains: "Some of the customers come and they say that we are not a real Belgian company or real French company and then they won't buy it. They want to see 'Made in Belgium' or 'Made in France' on the box. They want to give a gift and they don't want it to look like it's made in China. So we did a little research and talked about how we should go about making the chocolate and packaging. We decided not to do packaging toward Chinese tastes but do it like European packaging because chocolate is not a Chinese product. So that is why we never chose a Chinese name for the company. Customers think our name is better. And now they also know they cannot find our chocolate anywhere else."

Laurier and Polly also learned quickly that there are different types of customers in China: some who just go for the "glitz and glamour" and just want the big box and the packaging as a gift, and some who appreciate the chocolate and what they are doing. "They know we are not a mass-production company. They know we're trying to do the best quality possible and that's why they appreciate it," says Laurier. "That is not the perception of China anywhere. The perception is that the Chinese like mass production just like they produce mass production." In fact, to get closer to these customers who want to taste and explore for themselves and not just buy a gift, La Place plans to expand further into retail from the Internet. But when they do, Laurier and Polly will continue to listen to their customers both in terms of what kind of chocolate they want and what kind of packaging they want—a packaging version of what De Bondt did for the winery in Sicily—and then expand that to other customers as well. "They give us ideas of what they want and what the market may be interested in," says Laurier, "so we try to create something new out of that."

A process like this is, of course, only possible when the market is less saturated with competitors, as is the case in Costa Rica. In fact, George Soriano and Julio Fernandez of Sibú Chocolate are hoping to be *less* different in the future and create new businesses like their own and expand the market. They are willing to share everything they are doing in the hopes that they can inspire other chocolatiers to explore the possibilities of fine flavor but not simply imitate them. "Now that we have shown people what we can do with Costa Rican chocolate, we are willing to share that with other chocolate makers and chocolatiers so that we can help them source their own Costa Rican chocolate to make their bonbons," says George. "We want Central America and Costa Rica to be known for chocolates. We want people to say, 'Oh, Costa Rica—there is great coffee there and some great chocolate.' We've shown people that they can do it; now let's help them do it and give them what we did not have." That said, Julio adds, "we will always be a little step ahead of them but we will help them because their success means more success for us, too."

In Paris, Patrick Roger would admire this. He always shares his thoughts with manufacturers and chocolatiers from around the world. But the idea of listening to customers at all in the creative process as Laurier Dubeau and Polly Lo do is anathema to him.

Patrick believes process is not important to the customers—only the chocolate is. "The primary customer is *me*. I'm the one who controls their taste preferences. I am the dictator of taste. A customer is only unsatisfied if [a chocolate] is *not* well made. If someone makes a mistake it can upset *only* me. As soon as I know that a mistake has been made, that chocolate cannot appear in my shop. We can't fail. It's unacceptable. Customers are going to judge you the same way as when they buy a car or a house—with each bite, the customer is going to size you up. At my shop in Sceaux, there are 700 customers a day at Christmas who are buying my little chocolates. All the ingredients are listed, but this doesn't mean anything to them. I have the refined palate that I pass on to them. I am the one who must guarantee the quality of my work."

This devotion to providing customers with an experience that speaks to them and their desires is shared by most chocolatiers, especially those like Patrick who sell mostly to customers from their city. "Our clients are the people who live here. It's not like people just come to have a little bite of chocolate and then never come back," Patrick adds. "We are more than what you find in the big brand names in Paris. If a tourist buys a Fauchon chocolate, for example, in Tokyo or New York, he forgets about it. My shop doesn't work like that." In fact, Patrick now has his first international shop in what might be called ground zero for bonbons: in Brussels right next to the legendary Marcolini, and is one among 2,000 chocolate shops. But wherever he goes, Patrick expects the same reception: "In the future, I don't know if we will open shops in New York or Tokyo, but if we do, we will have the same clients there that we have in Paris. So there is no difference in clientele, no matter where the shop is. That is how it works. My clients couldn't care less about Fauchon chocolates once they have come to my store. Behind that business, they are all capitalists. They are all just financers. With us, it is a question of love and seduction."

Love and seduction are essential in markets where competition is fierce and imitators abound. Anne Weyns of Artisan du Chocolat in London understands what this means for her company as much as it does for a Patrick Roger. "I think what customers are looking for are products that reflect my sensibility and my inspiration," says Anne, which is why she expects her UFOs to remain a signature product for the future. "But everybody is sort of roughly offering the same thing and customers look for products and experiences that they cannot find anywhere else. Unfortunately, within the chocolate industry, because flavors and shapes and so on are fairly easily copied, people set up like Patrick Roger and just do something like him.

The number of people who have copied Patrick Roger's semicircle with very bright colors and caramel . . . could you not come up with anything else? It is a form of flattery, I know, but still. People seem happier to just pick ideas from the bonbons that work and mash them together. Unless you manage your own point of difference people will copy, especially anything that can be done very quickly."

Bart Van Cauwenberghe of De Zwarte Vos in Belgium sums it up this way: "I can't tell you how many times I've been to a restaurant and had goat cheese on the salad. Do you know how many cheeses there are in the world? Why is everybody working with the same goat cheese? How are you making people curious?"

Michael Recchiuti would agree but draws a contrast between inspiration and mimicry: "I started at a farmers' market and the farmers would come up to me and say, 'I've got three cases of lemon verbena that nobody wants. Can you do anything with them?' So I started exploring, doing the herbs and infused chocolates. I wasn't the first one. There were already people in Europe doing that and mushroom chocolates and all kinds of goofy stuff based on their region. If they are in Provence, they use lavender. In Brittany, salty caramels have been there forever as something you just put salt on. If they think that I created it, they are wrong. It is all inspired by someone else. We are all just interpreting. But what I've been seeing lately is people mimicking what I do and I want people to just kind of take their own direction."

And given the limited range of manufacturers providing chocolate—95 percent of chocolatiers buy their chocolate from chocolate manufacturers—differentiation can be difficult no matter the direction, especially when dealing with the well-educated consumers. According to Anne, most sophisticated customers in the United Kingdom are now so into chocolate that they have gone beyond considering cocoa content and even origin to the manufacturer to understand the differences: "If they see a 'Madagascar 64%' bar at a small chocolate shop, they know its probably Valrhona Manjari, so it is actually quite difficult to differentiate yourself."

Few chocolatiers, however, profess any intense desire or overarching plans for going bean to bonbon completely, if at all; it is just too expensive and not worth the trouble. Better to work closely with chocolate manufacturers who can do that work for you, and if and when you are big enough, even create custom blends.

Problem is, shops like Artisan du Chocolat are not big enough, so Anne found a mid-step that gives her a point of difference for the immediate future, at least: She works from "liquor" (unsweetened ground cocoa beans). "We don't take beans and we don't take finished chocolate," says Anne. "Producing from liquor widens us to a completely different supply base. We work with some large chocolate-producing companies that might have a couple of tons of really interesting liquor, which they would not make the most of by blending. For example, one company has a couple of tons of cocoa from Haiti and Vietnam and they think they can make better use of it than blending it. So they call me and say, 'We have a limited amount of this and do you want some?' It's quite an unusual situation. This has given us a clear point of difference and has widened our suppliers to include processing companies and cooperatives in local cocoa-growing countries."

This kind of difference can prove essential when there are not only several wonderful fine flavor chocolatiers competing for customers in the same area but also plenty more watching, ready to produce knockoffs. Thomas Haas can relate. His fresh fruit chocolates have been his signature product for years—a technique and style easily copied by others. So Thomas is always searching for different possibilities without sacrificing his integrity as a chef and artist.

Of course, innovating within the traditional bonbon form won't always work out. For example, at one point Thomas realized that all his chocolate bars were listed by percentages and the customers loved them. He made replicas of those in miniature chocolate squares with the cocoa percentages included on them so a customer could experience and understand the difference between a 75% and a 67%. The result? According to Thomas, "We got the whole line done and we printed in cocoa butter the tiny little number for the percentage and we got them all ready . . . and, oh my God, it did not sell. It was too boring. But the chocolate bar sells! It didn't make any sense. But that was the reality, and I don't want to be boring."

Boring is bad, but repetition is inevitable for chocolatiers who must balance their pursuit of new directions with the expectations of their customers. For example, the first products Michael Recchiuti released when he and his wife, Jacky, moved from Vermont to San Francisco—eight chocolates, sliced pears marinated in lime juice and coated in chocolate, and a bag of orange peel—are still around because they sell really well. Thomas Haas understands this well, which is why he says he is always changing but in increments: "It is a silent move. We have to be sensitive when we implement—both to customers and cost.

Say I have six new flavors, but I can't afford to have more than thirty flavors and I don't have space for all the new ones. If we take this, this, and this away? I guarantee you I will have so many people knocking on my door and complaining that I can't take it off because that's their favorite. So we are very sensitive when we say this is our standard program and we've got to keep this going so we make it a little more contemporary or a little bit finer and then slowly implement it. And sometimes we bite the bullet and take it away—something like the pear and replace it with peach. Hopefully that and throwing ideas around keeps the creative juices flowing."

Hopefully, indeed. Because as Charles Bukowski said . . .

"Bad taste creates many more millionaires than good taste"

In 2004, Mort Rosenblum opened the first chapter of his book *Chocolate: A Bittersweet Saga of Dark and Light* by describing a chocolatier dazzling his customers at his lone shop in a suburb of Paris. Later, we learn the chocolatier is scoping out his first Paris location. Today, Patrick Roger has eight shops, five in Paris. He still has that first shop in the suburbs—as well as his mammoth chocolate *laboratoire* complete with a sprawling ground floor and a giant office displaying his own paintings and African-inspired sculptures. How things have changed.

Just don't tell Patrick Roger *he's* changed.

"My mother always says, 'He hasn't changed. He's still just as annoying!'" Patrick says, imitating the voice of his mom. "Nothing has changed. Even when I opened in Paris. We just balanced out the pricing so that every chocolate had about the same price, even if they had different ingredients inside. We have more shops now, but my philosophy has still not changed. When I bought my first shop, I was the same way that I am now. We are who we are."

Okay, maybe "change" is the wrong word to use when describing Patrick Roger. Perhaps "evolved" is better. Whether that evolution involves the incorporation of new ingredients, new locations, or even new technologies, Patrick and his fellow chocolatiers remain who they are and taste is always paramount. But as we have seen throughout this book, many things have changed and are changing in the world of fine flavor chocolate.

As Patrick argues, pointing to changes coming to his laboratoire and the industry in general, "Technology will change but only as it leads to better taste—we must always be searching to create this orgasmic feeling. In fact, we will be able to create tastes that are more complex. Currently, we are only able to make the interior first and then the exterior around it, or we can make a mold and then put the interior inside. In the near future, we will be able to make both the interior and the exterior at the same time, or we will be able to make the exterior, and then the interior inside. We will practically be able to put a solid into a liquid."

But Patrick is concerned that all this may come at a cost beyond money: "What is going to happen is that we are going to push this technology to its breaking point. Soon there will be new technologies that will completely wreak havoc on the texture and the taste of chocolate. We are pretty much already at the point of ruining the texture of chocolate. In the future, the texture of chocolate will be a catastrophe."

Call that what you will, but that's evolution or change no one can believe in.

Nevertheless, most chocolatiers are optimistic as they try to keep pace and maintain a realistic view of their business. Consider what Laurier Dubeau says about his work at La Place Collection in Beijing: "We didn't go into this business to make money. If we wanted to make money we would just have a normal job somewhere else at a big company and make much more money than we do now. We do this because we like it. We like interaction with the customers, and we love to talk about chocolate with them and to see them appreciate the new flavors that they never associated with chocolate. This is a great experience and there is a big future for fine chocolate."

And it is a future that looks to include a growing number of women. At the manufacturing level, chocolate making is hands-on and all about using big machinery; women like Michelle Morgan of Zokoko or Chloé Doutre-Roussel remain exceptional. But that seems to be changing. At one recent Ecole Chocolat bean-to-bar class, all the registrants but one were female. On the chocolatier side, things have already changed. Maybe it's because the costs of opening a chocolate laboratory are high but much, much less than opening a chocolate factory. Maybe it's a natural evolution from the world of the pastry chef where women have been dominating the ranks in recent years. Maybe it's the small-scale appeal and the bonbon's existence at the intersection of art, cooking, and commerce. Whatever it is, women like Anne Weyns at Artisan du Chocolat are taking their places in the fine flavor chocolatier universe. Just look at the United States. Fran Bigelow of Fran's Chocolates in Seattle led the charge decades ago with her legendary Gold Bars and more recently gray salt caramels. A few notable women followed suit. In the past decade, dozens of female chocolatiers have made names for themselves across the country.

But Patrick Roger is right that some things defy change: the work it takes to create a fine flavor chocolate. "To make a chocolate, it must go through twenty or twenty-five manipulations—we have to touch it twenty or twenty-five times!" Patrick says. "My hard work brought me everything that I have today. I am someone who is constantly evolving and discovering things, based on where I am from. It's not simple. The most beautiful thing for me is still the smell of chocolate and, when it comes down to it, that is hard work.

"But in the future, for example in France, people don't want to work hard anymore. People are so lazy! The French don't feel like working. Economically, we have to find some solutions . . . it is impossible for me to abandon all the herbs, the zests, and the juices that I put into my chocolates. So in the future, the question for me is how will I be able to continue to work? How will I be able to cultivate in the gardens? These days, I tell myself that if I have the means to do my work, I am happy. The problem is that for me, the future is not a luxury. The only luxury in my work is the handiwork. In the end, it is a labor of love."

Michael Recchiuti shares the sentiment that chocolate is a really hard business. "It is grueling," he says. "It is physically demanding. There is a huge learning curve. The equipment is really expensive. The packaging is really expensive. Even if you have a ton of cash, can it be successful? I don't know. I think you really have to be passionate about it. I think one of the problems I've been having in the last eight years is that a lot of people approach me and really want me to tutor them and give them ideas about chocolate so they can start their own company. That's fine. People did that for me and so I'm trying to pay that forward. But then you get the people who just look at it as a widget. They have money and they have business savvy and have backers and they have never touched chocolate and have no intention of making it and they're just going to hire a French chef or something like that. That is really disappointing. You've got to be in it. Or just don't do it. You might make money but . . . at what cost?"

Thomas Haas waxes philosophical: "Growth . . . cannot be the only thing that drives you. If that is what drives me, then I have forgotten where I'm coming from. There is no way that I can be as good as we want to be if we multiply the business every year and just keep on growing, because we are not in the commodity business where we press a button and a T-shirt comes out. I am amazed at businesses that have those models. But no matter who they are, they lack flexibility and the sensitivity to details that we have. We believe that we are not going to get bigger, just better."

And better, not bigger, often means saying no or yes on one's own terms. For example, an airline recently came to Bart Van Cauwenberghe of De Zwarte Vos in Belgium and said, "Every customer that flies with us, we will give them two chocolates. Can you pack them separately?" Bart said no. He understood what they wanted and why they needed to do it, but he decided that's not what his brand was about. That, he felt, was for the big brands like Leonidas.

Or consider what happened when Recchiuti Confections went to the Fancy Food Show in January 2012, where in

Michael's words, "There is so much novelty jump-on-the-bandwagon chocolate it was very strange. We put together a very simple, funky booth and did no sampling. We decided it was going to be that if you know us, we can finally meet you and maybe you can come and see the factory. We are not after everyone, just that one or two percent that might have a store in Michigan that is super passionate about chocolate. We will sell to them. We are not going to sell to anyone else. My wife is stricter than me about that. If we have not heard of you, she will say, 'Let me see your store.' And then she will say, 'Nah' because they sell wind chimes and stuffed bears and chocolates in the corner. They are nice people and they mean well but it is not what we are connecting to with our brand. Do you really want to commit to selling your product to someone that is not your customer base?"

Thomas Haas once provided chocolates for 70 percent of the Four Seasons hotels and fifteen to twenty Ritz-Carlton hotels: 20,000 to 30,000 chocolates a week shipped to hotels from Maui to Texas to New York to Boston from his small chocolate kitchen in a light industrial area of Vancouver. But when his neighbors came knocking on the door, everything shifted. First, he sold to them informally out of the shop. Then the lines grew and on Valentine's Day and Christmas would extend around the block. Soon enough, Thomas had a decision to make: hotels or neighbors. He chose the local customers, built a factory in an up-and-coming area, and put a little boutique patisserie in the front, creating what Thomas calls "a chocolate shop with a little cappuccino bar." When that retail part of the business became a huge success, he dumped all his hotel clients except for Boston.

"We spend ten percent of our day saying no to ninety percent of the business that comes through the door, guaranteed," says Thomas. "'We would like to open a wholesale account.' 'We would like you to ship chocolate sauce.' 'We would like you to make this.' And then the restaurant calls and the hotel calls and we say, 'No, that's not what we are.' That's not who I am. I know the business works as it works."

And business for these fine flavor chocolatiers is crazy and messy and delicious just like the world of fine flavor chocolate as a whole. It is a tasty puzzle not unlike the constantly changing pieces of brown butcher paper Michael Recchiuti puts up on the walls of his factory, filling them with words and images of where he wants to go. "My staff says, 'Can we take them down?' And I say, 'No, I haven't gotten them fully out of my head yet. I leave the paper there and it kind of tortures me because I know the process isn't right yet."

But when the flavor is right, the process is in place, and the chocolate is ready, it's time for the three words everyone has been waiting for:

"Here, taste this."

CONCLUSION

We may find in the long run that tinned food is a deadlier weapon than the machine-gun.
—George Orwell

It is tempting when writing a book like this to end with some grand gesture—an inspirational "save the whales" kind of proclamation that is both wistful and hopeful. It is also tempting to go on and cover topics—from social media to ecotourism to expanding product lines and box and bar sizes—you could not address in more detail and ever-finer strokes.

We prefer to stop talking and take a more pragmatic, present-oriented, and yummy path and urge you to take whatever you learned from our words and the words of the people we spoke to and go out and explore, savor, and above all *enjoy* the world of fine flavor chocolate.

Slow down and really taste. Discover what you love (and don't). And no matter what or whom you choose, follow your heart, mind, and palate. As you taste, try to remember all that fine flavor chocolate connects to—from gene to bean to bar to bonbon. Remember the people and places and possibilities behind it. Remember that it is a food and not simply a commodity.

We've listed in the first appendix the names and websites of the manufacturers and chocolatiers who helped us put together this book. You might start your exploration with these companies, or choose one you've always meant to try or a local manufacturer or chocolatier you'd like to support.

Many of the manufacturers and chocolatiers we interviewed and countless others offer their chocolate directly to consumers or sell through upscale markets and specialty shops offline and online. Some manufacturers may be hard to find in certain countries or do not sell to consumers directly or ship internationally, but you can still explore what they offer for the next time you visit their cities. You can find lots of information on the Ecole Chocolat website: (http://ecolechocolat.com), or log on and download chocomap's Find Chocolate app (http://chocomap.com/) to find a place to satisfy your chocolate craving anywhere you are now and in the future.

And so we end where we began: To future generations, let there be the finest chocolate in the world.

INTERVIEWEES AND EXPERTS CONSULTED

Chocolate Experts (Alphabetical Order)

Mark Adriaenssens, R&D Director, Barry Callebaut,
http://www.barry-callebaut.com

Professor Angelo Agostoni, ICAM Cioccolatieri,
http://www.icamcioccolato.it

Christian Aschwanden, Max Felchlin AG,
http://www.felchlin.com

Shawn Askinosie, Askinosie Chocolate,
https://www.askinosie.com

Vanessa Barg, Gnosis Chocolate,
http://www.gnosischocolate.com

Brett Beach, Madécasse
http://www.madecasse.com

Richard Callebaut, Barry Callebaut,
http://www.barry-callebaut.com

David Castellan, Soma Chocolatemaker,
http://www.somachocolate.com

Martin Christy, Seventy%
http://www.seventypercent.com

Paul De Bondt, Cioccolato Artigianale de Bondt,
http://www.debondt.it

Steve De Vries, De Vries Chocolate,
http://www.devrieschocolate.com

Chloé Doutre-Roussel, Chloé Chocolat
http://www.chloe-chocolat.com
and El Ceibo Chocolate, http://elceibochocolate.com

Laurier Dubeau, La Place Collection,
http://www.laplacec.com

Carlos Eichenberger, Danta Chocolate,
http://www.dantachocolate.com

Julio Fernandez, Sibú Chocolate,
http://www.sibuchocolate.com

Philippe Givre, Valrhona École du Grand Chocolat,
http://www.valrhona.com

Guido Gobino, Guido Gobino Cioccolato Artigianale,
http://www.guidogobino.it

Francisco Gomez, Casa Luker,
http://www.lukeringredients.com

Mott Green, The Grenada Chocolate Company,
http://www.grenadachocolate.com

Gary Guittard, Guittard Chocolate Company,
http://www.guittard.com

Thomas Haas, Thomas Haas Chocolates,
http://www.thomashaas.com

Pierre Hermé, La Maison Pierre Hermé Paris,
http://www.pierreherme.com

Hugo Hermelink, Finmac Organic Cacao Plantation, Costa Rica

Frank Homann, Xoco Fine Chocolate Company,
http://www.xocogourmet.com

Felix Inderbitzin, Max Felchlin AG
http://www.felchlin.com

John Kehoe, TCHO Chocolate
http://tcho.com

Volker Lehmann, Director, Santa Cruz de la Sierra, Bolivia

Polly Lo, La Place Collection
http://www.laplacec.com

Rick Mast, Mast Brothers Chocolate
http://mastbrothers.com

Ramon Morató, Chocovic
http://www.chocovic.es and http://www.ramonmorato.com

Dean Morgan, Zokoko,
Michelle Morgan, Zokoko,
https://www.zokoko.com

Dan Pearson, Marañón Chocolate,
http://www.maranonchocolate.com

Santiago Peralta, Pacari Chocolate
http://pacarichocolate.com

Art Pollard, Amano Artisan Chocolate,
https://www.amanochocolate.com

Michael Recchiuti, Recchiuti Confections,
http://www.recchiuti.com

Jorge Redmond, Chocolates El Rey, Inc.,
http://www.chocolates-elrey.com

Patrick Roger, Patrick Roger
http://www.patrickroger.com

Frederick Schilling, Amma Chocolate,
http://www.ammachocolate.com.br and
Big Tree Farms, http://bigtreefarms.com

Ed Seguine, Choc Research Fellow, Mars,
http://www.mars.com

Duffy Sheardown, Duffy's Red Star Chocolate Ltd.,
http://redstarchocolate.co.uk

George Soriano, Sibú Chocolate,
http://www.sibuchocolate.com

Jeffrey Stern, Stern Chocolates,
http://jeffreygstern.com

Bart Van Cauwenberghe, De Zwarte Vos,
http://www.dezwartevos.be

Roger von Rotz, von Rotz,
http://www.echt-vonrotz.ch

Massimiliano (Max) Wax, Nazario Rizek CxA,
http://www.nazariorizek.com

Anne Weyns, Artisan du Chocolat,
https://www.artisanduchocolat.com

Joseph Whinney, Theo Chocolate,
http://www.theochocolate.com

Science Experts

Dr. Lyndel Meinhardt, Agricultural Research Service, US
Department of Agriculture (USDA-ARS),
http://www.ars.usda.gov

Robert Peck, World Cocoa Foundation,
http://www.worldcocoafoundation.org

Dr. Wilbert Phillips-Mora, Tropical Agricultural Research and
Higher Education Center (CATIE)
http://www.catie.ac.cr

Pham Hong Duc Phuoc, Director of the Biotechnology Center,
Nong Lam University, http://www.hcmuaf.edu.vn

Dr. Darin A. Sukha, Research Fellow, Cocoa Research Unit,
University of the West Indies, http://sta.uwi.edu/cru/index

HEIRLOOM CACAO
PRESERVATION INITIATIVE
FOUNDING CIRCLE MEMBERS

The following companies and people comprise the Founding Circle Members of the Fine Chocolate Industry Association's Heirloom Cacao Preservation Initiative.

Agostoni Chocolate

Amano Chocolate

Blommer Chocolate Company

Choclatique

Chocodiem

Chocolates El Rey

Cocoa Dolce Artisan Chocolate

Dandelion Chocolate

De Vries Chocolate

Dessert Professional

Ecole Chocolat

Flying Noir

Fran's Chocolates

Glacier Confection

Gnosis Chocolate

Guittard Chocolate

Manufacturing Confectioner

Marañón Chocolate

Mars Inc.

Mary Jo Stojak

Soma Chocolatemaker

SPAGnVOLA

Sweet Paradise Chocolatier

TCHO

the C-spot

(http://finechocolateindustry.org/hcp)

FINE CHOCOLATE GLOSSARY

As a founding member of the Fine Chocolate Industry Association (FCIA), Ecole Chocolat, with help from other FCIA members, volunteered to create the FCIA's Chocolate Glossary using general industry terms from the Ecole Chocolate Glossary. We've used those basic terms below with some updates, modifications, and additions. We hope this glossary helps you with any unfamiliar words in this book and beyond. For more terms and information on different types of bonbons and confections please go to the Fine Chocolate Industry Association page at: http://www.finechocolateindustry.org/confection-glossary.php.

Bonbon: An individual confectionery center, either enrobed with chocolate to cover or encased in a molded chocolate shell. Referred to as a "chocolate" in North America or "praline" in Belgium, Germany, and Switzerland. We use this term in the book instead of "chocolates" to delineate a confection rather than bar or bulk chocolate.

Cabosse: The French term used throughout the industry for the fruit of the cacao tree.

Cacao: Can refer to the Theobroma cacao tree, the fruit it produces, or the fruit's seeds. Once the cacao seeds are fermented and dried, they are then usually referred to as cocoa beans.

Chocolate: One of the end products of processing the seeds of a cacao tree. Other products include cocoa butter and cocoa powder.

Chocolate bloom: Also called "fat bloom," a thin whitish, beige, or gray film, streaks, or spots that form on the surface of chocolate as a result of many factors such as incomplete tempering; incorrect cooling; enrobing or molding cold centers; presence of other fats in the centers or chocolate; and storing the chocolate in too warm conditions. Does not impact the taste or condition of the chocolate but does mar the appearance and texture.

Chocolate maker: Refers to those companies producing chocolate from fermented and dried cocoa beans in small batches and in limited quantities. Usually their chocolate is sold as retail bars direct to the consumer or wholesaled to specialty retailers, with very limited bulk quantities available for use by chocolatiers in their bonbons or confections.

Chocolate manufacturer: Refers to those large companies that produce a broad range of chocolate from fermented and dried cocoa beans to supply wholesale to mass-market and/or specialty restaurants, patisseries, bakeries, hotels and chocolatiers. While historically chocolate manufacturers never sold direct to consumers, with the increased interest in fine chocolate, some traditional chocolate manufacturers now wholesale their own retail-size bars through specialty and mass-market retailers.

Chocolate liquor: Ground cocoa nibs, whether in molten liquid or solid block form. The term "chocolate liquor" has nothing to do with alcohol; it refers to the fact that during the grinding process, the cocoa butter is released from the nibs, rendering the mass into a "liquid" state.

Chocolate or cocoa percentage: Refers to the percentage of chocolate liquor plus any added cocoa butter, cocoa powder and/or cocoa nibs in a chocolate. The cocoa percentage has no bearing on the quality of the chocolate. For example, different 70% chocolates may range from excellent to terrible. The only specific thing that we can say with any certainty about a 70% dark chocolate bar, prior to tasting it, is that it has about 30% sugar in the formulation.

Chocolatier: This term usually refers to a person who sources fine chocolate produced by chocolate manufacturers or chocolate makers to create unique chocolate products, bonbons, and confectionery.

Coating chocolate or **chocolate-flavored coating:** Some or most of the cocoa butter is removed from the chocolate liquor and is replaced with less-expensive vegetable fat to produce an inexpensive product that does not require tempering. Also called compound coating, decorator's chocolate, confectioner's coating, and pâte à glacer.

Cocoa beans: Seeds of the cacao tree that have been fermented and dried.

Cocoa butter: Cocoa butter is rare among vegetable fats because it is mostly solid at room temperature, but starts to very noticeably soften and melt at just a few degrees beneath body temperature, leading to its unique melting mouth feel. These interesting qualities are due to the fact that cocoa butter is polymorphic, with six somewhat overlapping crystallization and melting ranges. Cocoa butter resists rancidity, and can be stored for much longer periods of time than most vegetable fats without spoilage. Additional uses include pharmaceutical and cosmetic purposes.

Cocoa nibs: The broken pieces of the fermented, dried, and usually roasted, cocoa bean, after the shell—actually the thin seed coat of the cocoa bean—has been removed via a process called winnowing. Cocoa nibs may be eaten out of hand or ground into chocolate liquor.

Cocoa powder: Once the cocoa butter has been hydraulically pressed from chocolate liquor, the remaining material is a compressed "cocoa cake." This cocoa cake is then reground and sifted until it is a fine cocoa powder. Cocoa powder, though lower in cocoa butter than the initial chocolate liquor from which it is made, will still have from 10 to 22% cocoa butter content as defined by the FDA.

Conching: A texture and flavor improvement process similar to kneading, that is carried out by any of a variety of machines called conches or refiner-conches. The process, which generally follows refining, takes place over several hours or days, depending upon the machine, the chocolate maker's vision regarding flavor and texture, and the particular cacao from which the chocolate is made. It is still not well understood what causes the significant flavor changes that occur within conched chocolate, though various food scientists throughout the twentieth century suggested that volatilization of certain flavor compounds, oxidation of others, and even the process of coating cocoa particles with cocoa butter, may play roles.

Confection: For the purposes of this book, refers to products that are not enrobed or shell molded, such as nut barks, turtles and clusters, dipped fruits and pretzels, molded lollipops or novelty items, and pastry or bakery recipes like macarons or brownies.

Couverture: The term usually refers to chocolate containing at least 32% cocoa butter that is made with concern for the overall flavor and texture of the chocolate. Couverture is generally used by chocolatiers to coat (enrobe) fillings or in the production of shell-molded bonbons, in which the fluidity of the chocolate is important in order to produce a thin shell or coating. Couverture comes from the French word *couvrir*, to coat or cover, and is pronounced koo-vehr-TYOOR. Sometimes referred to as fondant chocolate.

Dark chocolate: Fine dark chocolate should not contain any ingredients beyond chocolate liquor [could be referred to on the label as cocoa, cocoa powder, cocoa solids, or cacao], sugar, and cacao [cocoa] butter. Optional ingredients may include lecithin, and vanilla. Dark chocolate can be further labeled as bittersweet, semisweet, or sweet chocolate. These are arbitrary labels except in countries such as the United States, where use of the labels is regulated by the government depending on cocoa percentage.

Dutch process: The cocoa cakes, powder, or nibs are treated with an alkali salt to increase the pH value and neutralize the acidity. While this process doesn't change the flavor of the chocolate except for the acidity, it does deepen the cocoa powder's color, making it appear richer, and improves its suspension in liquids.

Enrobage: The thin, hard covering of a coated bonbon.

Enrobing: Coating an individual confectionery center in couverture to create a bonbon. Can be accomplished by hand or with the help of a machine.

Fermentary: The area set aside for fermentation of cocoa beans postharvest. This could be as simple as a cleared space in the forest or as complex as a series of boxes where cacao beans are rotated as they ferment.

Fondeurs: Chocolatiers are called *fondeurs* or "melters" in France, as they melt large blocks of chocolate and temper it to cover bonbons or mold into filled or solid shapes.

Ganache: The classic artisan bonbon center, an emulsion of chocolate and liquid (typically cream). Ganache can be flavored with fruit, nuts, spices, herbs, and aromatic liquids such as liquors or teas. Ganache is highly versatile and can be piped, slabbed, or shell-molded, but its most recognizable form is semi-round truffles.

Gianduja: A smooth blend of roasted hazelnuts or almonds with milk or dark chocolate, produced commercially and used for bonbons and desserts. This confection (pronounced gyan-DOO-ya) was created in the Piedmont region of Italy by Caffarel during a chocolate shortage in 1865.

Lecithin: Lecithin, an emulsifier, can be added during the conching process to improve fluidity and decrease the viscosity of chocolate. Some fine chocolate makers use lecithin while others do not; that is the personal choice of the chocolate maker.

Melangeur: A grinding machine that incorporates round granite wheels spinning in a circular motion to crush seeds and grains. In chocolate making it is sometimes used for the first grind of roasted and winnowed cocoa nibs.

Milk chocolate: Fine milk chocolate should not contain any ingredients beyond chocolate liquor [could be referred to on the label as cocoa powder, cocoa solids, or cacao], sugar, cacao [cocoa] butter, dry milk or cream solids, milk or cream fat. Optional ingredients may include lecithin and vanilla.

Pistoles: Originally this French word referred to gold coins in use in European countries until the late nineteenth century. Now, in the world of chocolate, pistole refers to coin-shaped pieces of couverture.

Praline: Refers to a bonbon of filled chocolate (in Belgium, Germany, and Switzerland); a New Orleans pecan sugar candy (United States); or a blend of chocolate with nuts, usually almonds (France).

Refining: This is the second grind of the chocolate liquor with ingredients such as sugar or vanilla. All of these chocolate recipe ingredients are then ground together so the particles are rendered the same size, producing a smooth, velvety texture.

Roasting: Cocoa beans are roasted to develop the characteristic aroma and taste of chocolate. The length of the roasting process and its temperature vary, though for those familiar with coffee roasting, cocoa roasting times and temperatures can generally be said to be significantly longer and lower. Fine chocolate manufacturers/makers generally do not roast every origin of cocoa beans in the same way, but try to find the combination of time and temperature that best enhances a particular origin's flavor.

Single origin: Cacao beans from a specific region, valley, farm, plantation, or orchard. There are no industry guidelines for the use of the term so it can mean any of the above.

Sugar Bloom: Water condensing on the surface is absorbed by the sugar in the chocolate producing white dots or tackiness. Difference between temperatures in storage and handling will cause surface condensation that precipitates sugar bloom.

Tempering: A process in which the temperature of the chocolate is manipulated for controlled crystallization of the cocoa butter to occur, thus allowing the cooled chocolate to have a good "snap," glossy sheen, and smooth mouth feel. Chocolate manufacturers/makers include recommended tempering specifications for each of the chocolates they sell, as each chocolate needs different temperatures depending on its ingredients and processing. In addition to technical knowledge, fine chocolatiers must develop a highly refined understanding of the tempering process through experience, because only this experience ensures that each chocolate product is perfectly tempered, even when automatic or semiautomatic tempering equipment is used. If real chocolate containing cocoa butter is melted for coating, enrobing, or molding techniques, it must be tempered before use to ensure it hardens properly. When tempered chocolate containing cocoa butter is then melted, it must be tempered again before use in coating, enrobing, or molding techniques to ensure it hardens properly.

Terroir: The French term *terroir* has been used in the wine industry for ages, and is also relevant when speaking of cacao. It refers to the various ways a particular place can have an impact on a given population of cacao, such as the effect of general and microclimates in the area, soil composition, and even the unique microbiology of the growing area and fermentary.

Truffle: A member of the ganache center family of bonbons. But the truffle has a unique history. Originally truffles were piped onto sheet pans, and once the centers hardened, they were shaped into balls and rolled in cocoa. Because of their uneven and rough surface, truffles were named after the real truffle, the fungus. Now truffle centers are hand formed, cut into squares and enrobed, or piped into round chocolate truffle shells. The advantage of these shells is that softer ganache centers may be used.

White chocolate: Fine white chocolate should not contain any ingredients beyond sugar, cacao [cocoa] butter, dry milk or cream solids, milk or cream fat. Optional ingredients include lecithin and vanilla.

Winnowing: Separating the shell of the cacao beans from the bean itself. The beans are first cracked to help release the shell from the beans. Then, a person by hand or using a winnowing machine uses airflow to blow the lighter shells off of the heavier bean pieces.

FERMENTATION AND DRYING
A Quick Summary of the Processes

Cocoa beans from the pod are intense in color—purple, intense white, or some shade in between; they look and taste nothing like chocolate. To get to the beans, which are really two cotyledons and an embryo inside a leaf and shell, the cacao tree fruits or pods must be harvested by hand and then opened. Each pod contains twenty to fifty beans individually wrapped in pulp and attached to a stem. The beans are separated from the pod and stem and then placed in containers. Farmers and plantation workers in the fields or jungles still largely do much of this work.

The beans are now ready for fermentation or what is called "en baba" in Central and South America.

Fermentation is a natural postharvest process that converts the sugars in raw cacao beans to alcohol, kills the germ, and develops the necessary elements that modify the composition of the beans so they will yield the characteristic flavor and aroma of chocolate during roasting. It is during fermentation that fine flavor cacao beans like Trinitario and Criollo start developing their flavors (bulk or ordinary cacao develop most of their flavor during roasting)—without it there will be little chocolate flavor. Depending on the country, fermentation takes place on the ground, in baskets, wooden boxes, or cylinders stored away from light. Depending on the varietal, the fermentation process lasts from two to eight days. The beans must be turned regularly to ensure even fermentation.

During fermentation, the beginnings of the taste we

associate with chocolate forms. While polyphenols go down during fermentation, your standard dark chocolate bar still has far more antioxidants than a glass of red wine.

Why ferment, and why ferment immediately after harvest? Because cacao is a fresh food, it can be host to a cadre of bacteria, fungi, and yeasts indigenous to the tropics once the pod is opened. Fermentation can't be done elsewhere lest the beans fall victim to the bacteria and their band of brothers. Kill them with the heat of fermentation or we risk our health, and some argue endanger our lives as much as flavor—a literal, and hence far more unpleasant, death by chocolate.

After the fermentation process, cocoa beans, as they are now called, have high moisture content. In order to be shipped or stored, they must be dried. This drying process differs depending on the climate or size of the farm, cooperative, or plantation. Cocoa beans can be dried out in the sun on trays, roads, concrete slabs or mats in the drier environments. In tropical areas where daily rainfall is the norm, beans can be dried in covered sheds and/or on tables over hot air. In either method, it is important that there is air circulating around the beans.

Once the moisture percentage in the cocoa beans has reached 6 to 7 percent, they are sorted and bagged. When it comes to fine flavor cocoa beans, the sorting process is very important because the cocoa beans are classified and sold in the industry by their size. After being sampled and approved by the bean buyer or broker, the bagged cocoa beans are then loaded on ships to be delivered to chocolate manufacturers.

NOTES ON RESOURCES

Writing a book on the future of fine flavor chocolate that includes no recipes, few actual products, and little history except for relevant background to the topics we covered is made much easier when there are already so many books, in-depth articles, and websites covering everything from chocolate's history to buying, making, and tasting chocolate. In fact, by the time this book is published, there will no doubt be even more new books and information available on and about fine flavor chocolate and chocolate in general.

The people we spoke to, however, wrote few of these books. While all of them have published articles, often appeared in the media and at events as guests or experts, had articles written about them, and/or blog constantly, only Michael Recchiuti (*Chocolate Obsession*), Chloé Doutre-Roussel (*The Chocolate Connoisseur*), Patrick Roger (*On a Quest for Chocolate*), and Pierre Hermé (several books, most recently *Macarons*) have written books on chocolate to date (the Mast Brothers' *Mast Brothers Chocolate: A Family Cookbook* is forthcoming); all of them have great information and/or recipes. For general history, there are many options, but start with the second edition of *The True History of Chocolate* by Sophie D. Coe and Michael D. Coe (2007). Mort Rosenblum's *Chocolate: A Bittersweet Saga of Dark and Light* (2006) and Maricel E. Presilla's revised *The New Taste of Chocolate: A Cultural & Natural History of Cacao with Recipes* have good information as well (and Maricel has a new book coming soon).

In terms of cookbooks, there are too many in addition to Michael's to even begin listing here. That they keep coming in many styles and languages is testament to the ongoing love of chocolate worldwide. Find one from a person or perspective that appeals to you and enjoy! We have compiled a list of popular books focused on chocolate on our Ecole Chocolate page: http://www.ecolechocolat.com/chocolate-books.php.)

For those of you interested in learning more about Rambam's Ladder (discussed in Part Two), check out Julie Salamon's *Rambam's Ladder*: *A Meditation on Generosity and Why It Is Necessary to Give* (2003).

Beyond books, you can find and track much of the latest news, statistics, science, surveys, and more—and all of the information, sources, articles, and studies cited in this book—posted and archived on the Ecole Chocolat News and Events page (http://www.ecolechocolat.com/news.php). Follow us on Twitter (@ecolechocolat) for the latest news and updates and join us on Facebook.com/ecolechocolat for more information and commentary. In addition to the Ecole Chocolat and FCIA websites, our interviewees' websites offer a wealth of information on and photos of their chocolate, their travels, how chocolate is made, where it comes from, the issues facing the industry, and more. These and other websites too numerous to mention are devoted in whole or in part to fine flavor chocolate and provide ongoing coverage of the industry in great depth.

As with fine flavor chocolate itself, we encourage you to explore all of them and more!

ACKNOWLEDGEMENTS

From Pam: I first want to acknowledge Jim Eber for his passionate voice in this book. In early 2011, I was looking for a professional writer who didn't know anything about chocolate—other than that he or she liked it—to work with me on a strange book about the future of fine flavor chocolate: the small segment of the chocolate industry focused on the people making the finest chocolate possible. I wanted this writer to bring a totally impartial point of view to the information we gathered—no preconceived notions about what chocolate is or isn't, how the industry works or doesn't. Jim jumped in taste buds first and his journey into chocolate over the year we spent conducting interviews and writing the manuscript has been a joy to behold. I'm grateful to Lisa Ekus for putting us together, and I think the story has been told much, much better and wiser because of Jim. A big thank you to my colleagues in the fine chocolate industry who willingly shared their thoughts, fears, hopes, and dreams. And a huge hug for Daryl who has always been my biggest fan, even when I came to him ten years ago with this crazy idea of starting a professional chocolate school. I couldn't have done this without you!

From Jim: The generosity, spirit, and talent of the amazing people who gave their time and in some cases opened their businesses and homes to be interviewed for this book is matched in every way by Pam; thank you for choosing me for this project and trusting me to help you realize your vision. Thank you, Lisa Ekus, for introducing us. Thanks to Chloé Doutre-Roussel for being such a willing victim of my incessant queries and Dan

Pearson and Steve De Vries for being the close readers Pam and I needed to keep us honest. Thanks to Michèle Bleuze, who hosted me in Paris and helped me overcome *mon mauvais français* by accompanying me to interview Patrick Roger. And, as always, thanks to Simon, for eating everything we put in front if him and smiling every time, and Amy, who, while more than thrilled to work for fine chocolate, is not only my guiding light in life but the most patient and best damned reader anyone could ever have look at a manuscript.

ABOUT THE AUTHORS

PAM WILLIAMS has been involved in the chocolate industry since 1981 when she created her first chocolate endeavor, the chocolate shop: au Chocolat. In 2003, she founded Ecole Chocolat Professional School of Chocolate Arts which delivers high quality educational programs to students all over the world. The Ecole Chocolat program curriculum offers what Pam considers to be the foundation for any successful chocolatier or chocolate maker—a complete understanding of chocolate chemistry and flavor analysis, recipe development, production techniques/equipment and business issues. Pam received the 2011 Fine Chocolate Industry Association's Recognition of Excellence in Service to the Industry.

As well as leading Ecole Chocolat, in 2006 Pam created chocomap.com to celebrate the chocolate industry and in 2011, launched a free mobile app, *Find Chocolate,* which delivers the chocomap directory of over 2500 chocolate shops to Android, iPhone, iPad and tablets. In her free time, Pam is passionately working on the FCIA's Heirloom Cacao Preservation Initiative (HCP) as a member of the HCP Executive Committee. Pam and her husband, Daryl, live in Vancouver, BC.

JIM EBER has nearly two decades experience in public relations, marketing, and collaborating on books and projects with personalities and brands across a variety of consumer and business areas. He has been directly involved in the launch of 14 *New York Times* bestsellers, served four times as a speaker/panelist at the International Association of Culinary

Professionals (IACP) annual conference, and even helped launch Virgin Cola and the Volvo C70 Convertible.

Beyond this book, Jim is currently working with the Fine Chocolate Industry Association on the launch of the Heirloom Cacao Preservation Initiative. His first cookbook collaboration, *Sweet Myrtle & Bitter Honey: The Mediterranean Flavors of Sardinia* by Efisio Farris (Rizzoli, 2007) was named one of the best cookbooks of the year by *The New York Times* and nominated for two cookbook awards by the IACP. His first of two business book collaborations with Jeffrey W. Hayzlett, *The Mirror Test: Is Your Business Really Breathing?* (Business Plus, 2010), was a *Wall Street Journal* and *USA Today* bestseller. He lives in Massachusetts with his wife and son.